DEATH IS ALL AROUND US

The Mexican Experience

William H. Beezley, series editor

DEATH IS ALL AROUND US

Corpses, Chaos, and Public Health in Porfirian Mexico City

JONATHAN M. WEBER

University of Nebraska Press
LINCOLN

Library of Congress Cataloging-in-Publication Data
Names: Weber, Jonathan (Jonathan M.), author.
Title: Death is all around us: corpses, chaos, and public health in Porfirian Mexico City, 1887–1913 / Jonathan M. Weber.
Description: Lincoln: University of Nebraska Press, [2019] | Series: The Mexican experience | Includes bibliographical references and index.
Identifiers: LCCN 2018032363
ISBN 9780803284661 (cloth: alk. paper)
ISBN 9781496213440 (pbk.: alk. paper)
ISBN 9781496214324 (epub)
ISBN 9781496214331 (mobi)
ISBN 9781496214348 (pdf)
Subjects: LCSH: Public health—Mexico—Mexico City—History. | Corpse removals—Mexico—Mexico City—History.
Classification: LCC RA452.M6 W43 2019
DDC 363.7/5097253—dc23 LC record available at https://lccn.loc.gov/2018032363

Set in Sabon Next LT Pro by E. Cuddy.

To Campbell Paige

CONTENTS

ILLUSTRATIONS

FIGURES

MAPS

ACKNOWLEDGMENTS

Writing a book is a wonderful but lonely exercise. However, I was lucky enough to have a supportive wife, family, friends, and mentors, both in Mexico and the United States, who made all the long hours bearable. I owe an enormous amount of debt to Dr. Ronald E. Doel and Dr. Kristine C. Harper, who have helped me grow not only as a historian but as a person and more importantly, a writer. I cannot say enough positive things about both Dr. Doel and Dr. Harper, and anyone who has had the opportunity to meet them and work with them instantly knows that they are both class acts and great role models for their students. In particular, both have helped shaped the book, never failing to provide detailed comments on my drafts that drastically improved the direction and quality of the final product.

I am also grateful to my parents and in-laws here in the United States and my faux parents in Mexico City. My mom and dad (and my in-laws, Stacy and Vits) supported my decision to pursue a graduate degree, the research that inspired this book. Moreover, the research for this book wouldn't have been possible without the wonderful generosity of Héctor and Glafira in Mexico City, who welcomed me into their home in 2005 when I visited for the first time to conduct research for my master's thesis. They treated me like one of their children, and without their help and kindness, I would never have been able to finish the project. I must also thank Professor Monica Rankin for graciously taking me under her wing, inviting me to writing sessions alongside her and a handful of other well-established scholars at the University of Texas at Dallas. Monica has also been instrumental in helping me shape the narrative and pushing me to reexamine the archival evidence to determine how I wanted to tell the story of dead bodies and technology in the Porfiriato. Additionally, I want to thank Professors Elaine Carey, Linda Arnold, Claudia Agostoni, and Ana María Carrillo for helping me conduct research and providing guidance over the years while I was in Mexico City. Linda, in particular, is a great person and historian whose countless suggestions and unmatched knowledge of archival holdings in Mexico City is a valuable resource for any young scholar, and I cannot thank her enough. The staff of the Archivo General de la Nación, Archivo Histórico de la Ciudad de México, Archivo Histórico de la Secretaría de Salubridad y Asistencia, and the Archivo Histórico de la Universidad Nacional Autónoma de México, thank you for your help in locating archival material (some in the official records catalog and some not), photography assistance, and overall kindness. Thank you to Jenn, who read through early versions of the manuscript and offered her suggestions; your help was much appreciated. I want to say a special thank you to everyone at the University of Nebraska Press, in particular, Bridget, Emily, Rosemary, Elizabeth, and I am sure countless others. In talking with Bridget early on in the project, she was genuinely excited about it, and her interest and belief in the project and her support is greatly appreciated. I am also thankful to Sarah C. Smith, my copyeditor, who has made the final product much stronger than the original. I was especially touched by her personalized notes scattered throughout the text, which just made me feel good when I read through the edits.

Lastly, I want to thank my wife for her patience over the years. Since my master's thesis, Kendra has stood by me, even when I wanted to quit several times. Truthfully, I do not know if I would have reached this point if she had not pushed me outside my comfort zone and taught me that if you really love something, you have to work at it. As I edit, during the World Cup, we've now seen three of them together and I'm aiming for at least twelve more with you. Thanks for being the best mom, partner, friend, and person I could ask for. You always believe in me, even when I doubt myself, and I am grateful for you. I love you.

DEATH IS ALL AROUND US

Introduction

According to the World Health Organization, a specialized agency of the United Nations concerned with international public health, as of 2016 the leading cause of death in Mexico is diabetes. Roughly 14 million Mexicans have diabetes, a number that has nearly tripled since 1990, and expert projections do not see the number slowing down anytime soon. In fact, epidemiologists project that by 2030 17 percent of Mexican adults will end up diabetic, and by 2050 half of all Mexican adults could suffer from diabetes at some point during their lifetime.[1] In previous decades, diabetes was a disease that only impacted 2–3 percent of the Mexican population, and the mortality rates associated with diabetes, 4.8 per 100,000 people in 1950, have increased sharply since 1990 as the death rate went from 30.8 per 100,000 in 1990 to 70.8 per 100,000 in 2014.[2] Epidemiologists like Dr. Tonatiuh Barrientos of Mexico's National Institute of Public Health believe that the rise in diabetes can be attributed to three factors: diet, green space, and physician naïveté. A prime culprit affecting diet, in his opinion, is soda, on which the government imposed a one-peso (ten-cent) tax per liter in 2014. To put the problem into perspective, according to Coca-Cola's filing in 2015, Mexico's annual consumption was six hundred eight-ounce servings per person per year, which is the equivalent to every Mexican citizen drinking two glasses of Coke per day. These numbers also do not take into account other soda brands, such as Pepsi or Pascual.[3]

Additionally, the lack of green spaces, available fitness equipment, and clear running spaces, especially in Mexico City, has exacerbated the public health problem. Heavy smog, cracked sidewalks, congested streets, and crime, as Barrientos puts it, have kept people from exercising, thus contributing to the rise in the number of diabetes cases. Barrientos and other medical experts also believe that for too long health officials put the responsibility of diet and exercise squarely on the broadening shoulders of their patients. As Barrientos sees it, and he is not alone, the government must intervene more directly when it comes to diet and exercise if Mexico's diabetes problem is going to improve.[4]

Experts further believe that Mexico's physicians are ill equipped to handle chronic disorders that require lifelong medical attention and monitoring, which requires a different skill set then what they are learning in medical school. Compared to infectious diseases, Dr. Gerry Eijkemans, the head of the World Health Organization in Mexico, noted, it is easier to reduce a mosquito population than to change a lifestyle, to change "the way a society is basically organized (to encourage) people to consume unhealthy food with lots of fat and sugar."[5] This attempt to shape the lifestyle choices of its citizens in the name of protecting their health is nothing new in Mexico. During the presidency of Porfirio Díaz (1876–80, 1884–1910), he and his state officials believed they had the ability to improve the lives of Mexicans through science and technology in order to solve Mexico's public health problems. But getting all citizens to adopt new ways of celebrating and mourning death would prove to be a Sisyphean effort.

Late nineteenth-century Mexico was a country rife with health problems. For example, in 1876 1 out of 19 people died prematurely in Mexico City, the country's capital, a city of 250,000 inhabitants. Compared to other world capitals at the time, such as London (1 out of 52), Paris (1 out of 44), or Madrid (1 out of 34), state officials realized, Mexico's capital city was one of the most unsanitary places in the Western world.[6] While the exact causes of death varied by individual, it is not an exaggeration to state that each day dozens of dead bodies, proof of such a staggering statistic, could be found scattered throughout the streets of Mexico City.

To combat such problems, President Díaz issued a decree in 1879 that the Superior Sanitation Council (Mexico's Board of Health) be responsible for supervising and guiding all public health policies and programs going forward. The goal was simple: to improve the health of citizens regardless of their cooperation. Díaz and his state officials wanted to turn Mexico into a modern nation and the residents of its capital, Mexico City, into hygienic and responsible citizens. Improving the health of Mexicans was a hallmark of how Mexican officials believed they would, as one observer noted, "drag the country out of the abyss, and place it, a well-organized, peaceful, and prosperous State, in front of the civilised Powers of the world."[7] Evidence of the push to modernize the cityscape of the capital appeared in many forms. For example, photographs and guidebooks from the early twentieth century emphasized improvements such as wider streets, parks, or street lamps, all efforts to provide a safer environment. But underneath the improved safety of these new features, state officials wanted to replicate the physical changes that had swept through Paris during the presidency of Napoleon III in the mid- to late nineteenth century. As Paris had, Mexico City created manicured green spaces, visually appealing buildings made from iron, new sewage systems that could handle both rain and increasing levels of human and animal waste. The physical changes to the urban landscape served as visual proof to city residents that the city was changing for the better. To make room for the future of Mexico, state officials had to tear down the narrow, dark streets and dilapidated buildings in parts of the capital that were constant reminders of Mexico's past.

Porfirian officials viewed the changes underway in Mexico City as an important step in their plans to shape the population. They considered their efforts similar to those of a gardener who trains, prunes, plants, and weeds out unwanted or undesirable plants from their gardens. In doing so, the gardener imposes his or her own "principles of order, utility, and beauty on nature," and the result is a garden that grows a "small, consciously selected sample."[8] But what grew in the city alongside improvements in public health and technology, enjoyed by both upper and middle-class citizens, was a growing discontent among the lower classes who resented having their lives shaped by a government that did not value them or their contributions.

The changes that state officials introduced in the city targeted the urban poor, who resided in and around all areas of the city deemed worthless. Removing them from the landscape forced the poor to seek shelter elsewhere, usually in growing neighborhoods that popped up on the outskirts of town and, at such a distance, lengthened both their commute and work day. Such burdens placed on the daily lives of these individuals by the rapidly modernizing environment resulted in conflict between the desires of state officials and lower-class citizens. State officials believed improvements to the landscape would not only beautify the city but also improve public health and the behavior of residents, especially those from the lower class, whom officials considered unrefined and barbaric. Yet these residents often ignored how state officials wanted them to act or use the new modern spaces and behaved in ways that undermined the state's efforts in social engineering.

Nevertheless, the Mexican government's goal to shape the behavior and attitudes of the Mexican people remained a central tenet of the administrations that followed the Porfiriato. Even a civil war in the country, the Mexican Revolution, failed to change much of the vision that Díaz held for transforming the social order. While new faces battled over the direction of the nation, controlling behavior and attitudes of citizens, especially the poor, remained an integral feature of their agendas. State officials continued to target behaviors they considered a threat to modernity: drinking, gambling, prostitution, sex, and death.[9] Nowhere did this threat appear more than in state institutions like public cemeteries. These were spaces where the behavior of workers, usually from the lower classes, continued to represent one of the greatest paradoxes in Mexico's attempt to transform its population.

In November 1916 two cemetery laborers, José Espinosa (fifty years old) and Marcelino Rosas (forty-five years old), stood before a judge in Azcapotzalco, a suburban town located in northwestern Mexico City. The charge, profanation of human remains, was a crime that ranged from simple exhumation to the dismemberment of a dead body or even necrophilia. However, for Espinosa and Rosas, the profanation in question was the illegal exhumation and reburial of corpses in exchange for money in the local cemetery. According to their testimony, both Espinosa and Rosas worked as gravediggers in Panteón de San Miguel Amantla. While at work a year earlier, in November 1915, a friend of Espinosa's named Juane Telles, the party who would later accuse the two of profanation, showed up at the cemetery to ask his friend if he could help him out. His son had died recently, and all the family could afford was a burial in a fourth-class plot. To help Telles, both Espinosa and Rosas offered to bury his son in a better location, "a first-class grave," their testimony revealed, in exchange for ten silver pesos.[10]

Telles agreed to their proposal. After receiving payment, Espinosa and Rosas moved the boy's corpse to a first-class grave, but they later moved it to a second-class grave where "it has been to this day." After telling the police about what had occurred at the cemetery, both men lamented their actions. According to Espinosa, he "took the money because his salary as a gravedigger was not enough to support his family," and, according to Rosas, "he did not think what they did was a crime." Despite having committed a crime, their attorney, José Camara Arjona, made sure to depict his clients not as criminal masterminds but men of "good conduct." Camara Arjona had multiple character witness testimonies that pointed out that Espinosa's only major flaw was that he was "rude and illiterate," but this was not behavior of a hardened criminal.[11] According to Camara Arjona, Rosas only helped

Espinosa because he had hired him and thus felt indebted to his friend. Camara Arjona hoped the judge would be sympathetic to their situation. These men were not criminal masterminds but down-on-their-luck individuals who had seen and taken an opportunity, albeit a poor one, to improve their economic situation, which was an admirable quality.

But before the judge could make his ruling, the prosecution presented the testimony of a witness who told a different story about the true nature of the defendants. Luis Guerrero Ramírez told police that he had visited the cemetery where the duo worked to inquire about the possibilities of burying his deceased child. However, upon his arrival, he found the cemetery closed. But before leaving, he was able to find an employee, José Espinosa, who told him that while space for additional burials was sparse, "more could be formed" if he and his team, a total of four men, received a tip in the "amount of 30 pesos plus 10 pesos per man." Once this additional information came to light, the judge decided to issue his verdict: guilty. In addition to the judge deciding "to admonish both men for their crime," he sentenced both to "two months in jail and a fine of 20 pesos, which if they are unable to pay, would result in an additional 16 days in jail."[12] While both Espinosa and Rosas ignored the law in order to supplement their meager incomes, their actions violated the good faith that state officials placed in state-employed workers, whom officials believed would ascribe to a definition of honorable behavior befitting a modern citizen.

Low-ranking workers from state institutions, however, often chose to engage with expected behavior and official rules in ways that fit into their moral economies. Much of this had to do with the reality of day-to-day survival and the effort to provide food for their families, even if their actions could result in potential jail time. Espinosa's and Rosas's reliance on the state-supported forms of patronage and assistance had not provided enough to satisfy their needs and wants. Instead, opportunities arose based on their specialized form of knowledge and experience in the cemetery. Burying corpses in different areas of the cemetery did not violate any of the rules that they had created for themselves. In fact, in their opinion, their approach to burials was less tedious since it eliminated the required paperwork, which accelerated the burial process and thus reduced any potential threat to public health posed by corpses. Nevertheless, the moral economy surrounding death that lower-class workers had created represented a true threat to the Mexican state officials' goals for the modern state.[13]

Circumventing the official rules surrounding death reflected how the values of lower-class citizens—primarily, taking care of one's family by any means necessary—clashed with the version of modernity promoted by state officials, which valued obedience to the rules and the state itself. Lower-class citizens, unfortunately for the state, were not pliable, inanimate objects. Instead, they were people with feelings, emotions, needs, and wants, which state officials believed they could erase by tapping into these citizens' desire to belong and support something greater than themselves: the modern Mexican state.

But this attempt to cultivate a modern image for Mexico that focused on improving the behavior of citizens, their hygiene, and other matters related to the body was nothing new for Mexicans. Much of this desire to control the population can be traced back to the arrival of the Spanish in the Americas in the early sixteenth century. When the Spanish gained control over Mexico in the early sixteenth century (after Hernán Cortés's conquest from 1519 to 1521), several important institutions emerged that reflected the monarchy's desire to control the local population, the most important of which was the Catholic Church. The church was instrumental in defining medical practices in the New World, including who indigenous healers known as curanderos could treat, the legality of medical procedures and treatments, and the role religious ritual should play in an individual's health.[14] However, while some of the population embraced the changes begun by the church, especially those inhabitants with ties to Spain, the remaining indigenous population rejected much of the Catholic Church's changes, opting instead to pick and choose elements that fit their worldviews.

Nevertheless, the Catholic Church remained committed in its desire to control the behavior and lives of the inhabitants of New Spain. In particular, church officials focused on matters related to death, especially how best to remember the dead and prevent them from remaining in a state of purgatory, trapped between the worlds of the living and dead. Bishop Bartolomé García Jiménez of the Canary Islands (a Spanish possession) explained that some deaths resulted in the deceased's soul traveling to a purgatory rather than on a straight path to heaven, which meant that the soul needed additional help in order to rescue it from limbo. In return for their help, Spaniards and Spanish subjects would receive protection from the deceased once they reached heaven, pledging to protect both Spaniards and Spanish subjects in heaven or accelerate their ascent from purgatory into heaven upon their deaths. Additionally, these souls asked the living to celebrate their memories in banquet by devoting one day a year to their remembrance. García Jiménez's testimony led Spanish King Charles II to issue a royal decree that all his subjects needed to join him in prayers to remember and honor the souls in purgatory every year on his birthday, May 29. He also instructed priests to encourage their parishioners to choose an additional day so that they could pray for the souls in a more personal manner.[15] Both the testimony and the royal decree were essential components for demonstrating how the Spanish monarchy and Catholic Church sought to create an intimate connection between the living and the dead, which was not only for the good of the individual but for the whole of the Spanish Empire.

While the church was developing this connection, citizens began to form social organizations that created community allegiance by offering ecclesiastical insurance, as well as opportunities for social mobility and participation in the democratic process. These religiously inspired groups encouraged conversion but did not require uncompromising acceptance of the faith. Nevertheless, a significant percentage of their efforts went toward establishing a solid foundation for its members upon their deaths: for example, making sure the deceased could receive a proper Catholic burial, which included the corpse either beneath the church floor or in a space close to the church. Moreover, in burial ceremonies, the church made sure to place the appropriate religious paraphernalia (pictures of saints, crosses, and rosary beads) with the corpse, as well as giving a solemn remembrance through mass and prayer.[16] Members donated their own money, clothing, food, and religious iconography to help cover the cost of a "proper burial," thereby ensuring the deceased a place in the afterlife.

However, both the Catholic Church and government authorities grew weary of these citizen-organized groups as their number grew. For example, Mexico City boasted of three hundred such *cofradías* as early as 1585.[17] Threatened by a loss of power as cofradías flourished, church and state officials worked together to regain control over festivities that they had once dominated, especially those surrounding death. In particular, celebrations like that for the Day of the Dead often involved alcohol and led to alarming behavior, and officials believed these celebrations fostered drunkenness, sacrilege, and disorder. Church and state officials believed that this behavior was a threat to their power and control over death. Thus, they codified penalties for such behavior ranging from prison time to hard labor.[18] The result was the beginning of a long history of confrontation between the Catholic Church, state officials, and a local popular culture built on the domestication and popularization of a death cult.[19] How the individual citizen chose to celebrate, remember, and mourn the dead would become the nexus between the modern state and citizenship.

On the eve of Mexican independence in the early nineteenth century, the control the powerful Catholic Church had held over death for centuries was starting to unravel from within. While the church struggled with civil authorities over who had the final say over an individual's death, a schism occurred in the church hierarchy. On one side stood a group known as *sensatos*, whose piety centered on the individual and internal discipline. This group believed the church had monopolized the process surrounding death, most importantly burial rites and the steps required to reach the afterlife.[20] On the other side were those who believed that the individual needed the Catholic Church, along with its priests, to serve as their intermediaries to the divine.[21]

At the same time, however, the majority of the Mexican population did not share these views. Rather than operating at either end of this spectrum, the majority of people displayed a hybrid form of piety, much to the dismay of church officials and state officials later on. For example, last wills and testimonies often only included an unspecified burial location in a local parish, which contrasted sharply with burials in earlier years when location of their eternal rest was of upmost importance for burials because it served as a measure of an individual's piety. Closer to the church (better yet, inside it) meant that an individual was more pious and held greater value in the eyes of God. Additionally, ostentatious funerals from earlier periods, which had included caskets made from expensive and rare materials, began to disappear as individuals chose modest burials with simple, unadorned caskets.[22] Less was more during this transition period.

Ambivalence toward burials centered on the contested definition of piety. For years, the well-to-do had used their economic success to receive burials beneath church floorboards, which demonstrated to everyone their level of piousness. Burial beneath the floorboards and close to the pulpit meant that not only did they have money while alive, but even in death they were more successful than others. However, over time, this approach to burials further exacerbated class, racial, and religious tensions in Mexico. These burials also represented a more tangible problem for attendees because of the decomposition occurring under their feet, which produced pungent odors that filled the church, especially during warm summer months. The result was a growing fear that these odors would not only make parishioners ill but could cause debilitating sickness, which could lead to death.

As a result, enlightened reformers suggested that burials take place in cemeteries in the outskirts of cities rather than beneath church floors, which they believed would eliminate sickness and any threat posed to public health by the deceased.[23] Instrumental in changing the burials was King Carlos III of Spain (House Bourbon), who in 1787 issued a royal edict that sought to limit the spread of disease that officials believed corpses caused. Colonial officials and medical authorities began to require the construction of additional cemeteries at the margins of cities to protect public health and attempt to diminish the role class had played in burials in previous years: often local elites were the individuals who buried their dead underneath church floors, crypts, or in graves on church grounds. Additional royal edicts appeared in 1804, for example, that sought to reduce the influence of the Catholic Church in burials by prohibiting burials underneath patios or church courtyards. People had chosen to bury bodies in these locations because local church officials had emphasized the importance of burying the dead near the presence of saints, who some parishioners believed would watch over the deceased. These late colonial changes to burials initiated by King Carlos III, as historian Heather McCrea has shown, tried to "tighten the grip of the crown on its Latin American colonies" over public health and civil disorder, especially when it came to death as a cultural and religious practice.[24]

These burial reforms, however, would soon become part of larger social and cultural movements in the early nineteenth century that attempted to secularize Mexico. By the 1830s, Mexican liberal intellectuals like José María Luis Mora led anticlerical attacks against the Catholic Church and its "abuses of superstition," including the centuries-long monopoly the Catholic Church had held over burials.[25] To create a new image of Mexico, new laws emerged in 1855 and 1856 respectively that sought to reduce the power of the church in a newly independent Mexico. In 1855 Ley Juárez removed the church's judicial privilege (*fuero*), which had exempted it and all its personnel from civil and criminal prosecution in secular courts. Afterwards, in 1856, state officials instituted Ley Lerdo, which prohibited institutional ownership of properties and aimed to reduce the Catholic Church's vast rural and urban landholdings so that the state could sell them at auction to raise money for its own coffers.[26] The belief among liberal leaders was that the elimination of these privileges would allow the secular state to control marriage, baptism, last rites, and most importantly death. All would yield an incredible amount of money for the government. Furthermore, the Mexican government's control over burials would present state officials with an opportunity to implement modern healthcare policies that could improve the hygienic practice of citizens. Rather than superstition or tradition serving as the backbone of death and the burial process, as liberals had accused the Catholic Church of supporting, the burial process would be grounded in medical science.

But the shift from church-controlled death to state-controlled death made the entire process far more clinical for families. The result was that family members found it difficult to mourn and celebrate in traditional ways, which centered on the amount of time the corpse remained unburied. Indeed, it was customary for impoverished citizens, especially from indigenous backgrounds, to mourn the dead for more than a week. According to some firsthand accounts, when a person died, he or she was taken from the bed and placed on *petate* (a bedroll made from the fibers of the palm of petate), covered with a sheet, and lifted onto a table. Next to the corpse, newspaper was spread out, where the family made a cross from sand and lime. Additionally, family members would also wash and iron all of the clothes of the deceased before laying them next to the body. A candle burned day and night for nine days, during which there was a continuous wake for the deceased, and on the ninth night, visitors and family members brought tamales, mole verde, oranges, chocolate, and bread, which they left for twelve hours on the house altar to provide the deceased with food for each month of the year.[27] Afterward, the body would usually be taken to a local cemetery (or at least near one) or occasionally be dropped off next to shrines for patron saints.

This approach to death remained part of the Mexican culture

that President Porfirio Díaz sought to eliminate in the name of modernity. Thus, to ease the transition to modern public health, President Díaz, who ruled from 1884 to 1910, relied on similar burial policies established in the Bourbon era as the foundation for his modernization of Mexico. Yet the rising costs of burials, especially for the urban poor in Mexico City, meant that if the family desired a "proper burial" in a cemetery, the only remaining option became the free burials given with corpse disposal at designated deposits throughout the city. However, the urban poor used these deposits so frequently that they began to overflow, and corpses spilled onto the streets, where they became public health hazards. This result was not what state officials had intended with their construction, and the overflow represented a clear challenge to how state officials believed their modern capital should appear: an environment free from decomposing corpses.

Additionally, the capital struggled to keep its living residents healthy. To improve this situation, President Porfirio Díaz issued a decree in 1879 that made the Superior Sanitation Council (Mexico's Board of Health) responsible for supervising and guiding all public health policies and programs. His goals were simple. He wanted to improve both individual health and public health in order to transform the capital into something befitting of western Europe, and its residents into hygienic and responsible citizens. Thus, improving individual health would become the hallmark of the plan President Díaz had implemented, as one observer noted, to "drag their land from oblivion, its resources from bankruptcy, teach outlaws peace—in fact, make a nation and a prosperous country out of chaos."[28] To achieve such lofty goals, change had to start with those who comprised the majority of the population: the lower classes.

Thus, Porfirian officials created a discourse that linked death, public health, medical science, and technology into a cohesive narrative that promoted Mexico City as a model of modernity for the rest of the country. Central to achieving this goal were the various population management techniques employed by state officials that combined reason, science, and technology to organize and manage citizens' everyday existence in life and in death. Like the government in the United States during the nineteenth century, the Porfirian state "idealized the passive and submissive citizen—one who accepts that the state operates in a benevolent manner."[29] An integral component of this approach was the way the state celebrated dead heroes and important state officials. Dead bodies became not only "useful and effective symbols for revising the past" but the future as well, helping Díaz claim political legitimacy over the nation.[30] Moreover, state officials began to use these dead bodies as way to create a bond that tied death, the state, and citizens together. State officials carefully chose to put words into the mouths of the dead in order to support the changes they had made to public health policies, including the management of death.[31] For late nineteenth- and early twentieth-century Mexico City, the dead spoke loudly about the problems surrounding public health in the capital. Modern science and technology presented Mexican state officials with an opportunity to eliminate these complaints as well as the persistent stench of death on city streets and inside homes—or so they believed.

Public Health and the Modern State: Reappraising the History of Modern Mexico

For historians, the role played by state officials in matters related to public health has shifted over the course of the twentieth century. While groundbreaking historians such as George Rosen in the 1950s saw the modern state as altruistic when it came to the health and welfare of citizens, changing notions about the true nature of this relationship, especially during the late 1960s and 1970s, ushered in a new type of history that sought to challenge such a positive portrayal of those in power.[32] Such was the case with a bold thesis introduced by historian Thomas McKeown, who argued that while modern medicine aided by state officials had improved the individual's immunity to major diseases, individual and public health had actually improved during the eighteenth and nineteenth centuries largely due to the adoption of better hygienic practices, standards of living, and dietary changes.[33]

McKeown had assumed, however, that such changes were equal across social classes, which simply was not true. While his work opened the door for others to begin to cast doubt on the true nature of the modern state and public health, the controversial French historian, sociologist, and philosopher Michel Foucault ripped it off its hinges. Foucault's work challenged scholars to rethink the nature of the relationship that existed between the modern state and its citizens. He argued that the approach to governing citizens had relied on creating a distinct relationship between power and knowledge. According to Foucault, modern Western governments had developed new mechanisms of citizen surveillance over the course of the eighteenth and nineteenth centuries that revolved around the concept of discipline. The introduction of new professional fields of study such as criminology, psychiatry, medicine, and sexuality—all of which, as Foucault explained, had been forged by words and things—led government officials to gain the authority *to see* and *to say* what was wrong in society.[34] Thus, state officials were able to sort society by class and race in order to categorize sicknesses into such groups as criminal, mentally insane, or homosexual. State officials associated all of these labels with lower-class citizens, which allowed medical professionals, on behalf of the state, to use their professional expertise to present treatment plans to remedy individual and public health.

Despite his innovative thesis concerning public health and the modern state, Foucault's work was not without its weaknesses. The biggest problem scholars had with Foucault's work was the fact that he neglected to discuss how citizens could negotiate or challenge government-imposed techniques of control. However, anthropologist James C. Scott's work on the everyday experiences of peasant farmers in Malaysia helped fill this gap. While not a study of public health, it did combine Foucault's contempt for the policies of the modern state with an exploration of how poor citizens resisted state-driven reforms. As Scott argued, most lower-class citizens throughout history had not been interested in changing the larger structures of the state or the law but rather in exerting their autonomy through ordinary measures such as "foot dragging, dissimulation, desertion, false compliance, pilfering, feigned ignorance, slander, arson, sabotage, and so on."[35] Together, these two have influenced several generations of historians of Latin America who have used their work to better understand the relationship between citizens and state officials, especially in matters related to crime, death, and public health.

In order to better understand the relationship between the government and citizens that existed during the Porfirian era, it is important to recognize how the modernization process unfolded in Mexico. In late nineteenth-century and early twentieth-century Mexico City, this was at best an uneven development. Porfirian officials sought to make the city safer, cleaner, healthier, and efficient, all of which would provide a better image of the capital at home and abroad. Moreover, the push to turn citizens into something more was part of the larger desire by state officials to more simply organize the population. This was part of the larger agenda of modern statecraft that sought "to reduce the chaotic, disorderly, constantly changing reality" of society in favor of a process known as internal colonization. It was better that people became homogenized, largely through methods that shaped their behaviors and customs, which in turn made society "more amenable to the techniques of the State."[36] If successful, this would reduce the unpredictability that state officials associated with a heterogeneous citizenry. However, the ability to control the behavior of citizens in the name of modernity was an impossible feat. State officials who considered themselves modern based on the employment of economic rationality, scientific truth, and technological efficiency, French sociologist Bruno Latour argues, believed that nature and society were concepts they could separate and manipulate in their favor. However, nature and society were not separable concepts. Those who believed that they had successfully managed to achieve separation, Latour concluded, had been blinded by desire. What they failed to understand was that nature and society have always been interconnected, and thus there has never been a point in history when people have been able to achieve modernity.[37]

The most important studies that address this issue in Porfirian Mexico have examined how crime, death, and public health were all integral to the vision state officials had to manage and transform Mexican society. In the area of crime, the two most important works on Mexico have each addressed how state officials and elite members of society constructed a set of characteristics for identifying criminals or those most likely to commit a crime in the future. Not surprisingly, these characteristics focused on the innate deviancy of members of the lower classes, whose behavior and customs were obstacles to the progress that state officials wanted to achieve. Historian Pablo Piccato has argued that by committing crimes, the urban poor asserted their autonomy and challenged the modernization reforms implemented by the Porfirian government. Those in positions of power, ranging from police officers to teachers to cemetery administrators, sought to control the behavior of citizens, especially the poor, by admonishing how they dressed, what they ate and drank, or how they celebrated or mourned, among other behaviors. The urban poor instead chose to ignore them and act in ways that officials continued to classify as inappropriate, backward, and unbecoming of the modern Mexican.[38]

Additionally, the overwhelming presence of prostitution in the capital was a public health problem that required state intervention. As historian Katherine E. Bliss has shown, Porfirian officials believed that the behavior of prostitutes was a major obstacle to achieving modernity in the capital. But rather than banning prostitution altogether, state officials and medical authorities worked closely to develop regulations that would allow for state-sponsored monitoring. Prostitution as an industry was considered a necessary evil but one that Porfirian officials believed they could shape through the application of uniform guidelines and scientific protocol. For example, new regulations required prostitutes to submit to frequent and invasive medical exams in an effort to reduce the spread of disease. Yet these rules only applied to the prostitutes and not their male clientele, who also carried contagious diseases that they often brought back to their families. In spite of these state-driven efforts, as Bliss explains, Mexico City still had thousands of prostitutes who remained unregistered and operated outside the framework established by the Porfirian state. Prostitutes began to challenge the official narrative that they were broken, disease-ridden, alcoholic, and melancholic creatures who needed the paternal assistance of the state. Instead, they created their own definitions of honor, love, respect, friendship, and citizenship. Furthermore, following the Mexican Revolution, prostitutes began to argue that as part of the revolutionary government's social reform they too should be included in government welfare campaigns since they were important citizens who deserved the respect and support afforded to other segments of society.[39]

While regulating the bodies of the living was a central tenet of the Porfirian modernization campaign, so too was the management of the dead. State officials sought to define what constituted a proper and modern death in order to better facilitate their control over the lives of citizens. Anthropologist Claudio Lomnitz has explained that death in Mexico remains a complex idea that intersects with many avenues of society, from *corridos* (popular ballads) to the government's registration of death (burial registration and death certificates). Nevertheless, there has been one constant. Beginning in the Porfiriato, families had to go through official bureaucratic channels in order to have a relative buried, which, he has argued, gave state officials control over the death of citizens, as it allowed them to keep detailed records, commodify death, and construct a narrative of death that demonstrated to citizens who was in control.[40]

To further demonstrate this control, Porfirian state officials used the extravagant funerals of important state officials as a method to illustrate state power. Historian Matthew Esposito has demonstrated that over the course of Porfirio Díaz's presidency, he and his officials constructed dozens of national monuments, performed countless commemorations, and held more state funerals (110) than in any other period in Mexico history. All of this was part of Díaz's attempt to demonstrate to citizens that the government was in control of the bodies of citizens. Esposito has argued these funerals allowed the state to construct a past, present, and future vision of the national community for citizens, which was where state officials sought to instill the customs of modern cosmopolitan life to the city's growing population, many of whom had emigrated from rural areas.[41] This resulted in an attempt to continue to extend control over the bodies of citizens by creating uniform methods for celebrating death that reflected the principles of modernity, including burial reforms at city cemeteries that conformed to the state's vision of modernity.[42]

Controlling crime and death was part of the state's larger agenda to improve public health in the city and throughout Mexico. The two most important works to address this issue were written by Mexican historians, Ana María Carrillo and Claudia Agostoni. Each has traced the history of modern public health in Mexico during the nineteenth century and how the Porfiriato ushered in a new era of autocracy for health, tied to state officials' push to bring change to the country in the name of modernity. Public health first became a major problem for state officials in the late eighteenth century, when the Spanish Bourbon monarchy had attempted to modernize the residents in New Spain through the use of science.[43] The reforms instituted by the Bourbon kings evolved over the year as the king sent representatives to the New World to report back about the health of citizens. This led to the establishment of both sanitary and medical discourses that sought to create healthier citizens and regulate public environments with official health boards. Once the Bourbon kings lost power in New Spain, after Mexico's independence movement from 1810 to 1821, new approaches to improving public health appeared. However, those involved in this process wanted to create their own Mexican version of public health and medicine instead of one created by foreigners such as the Spaniards. Historian Ana María Carrillo has argued that after independence the Mexican medical community started a campaign to make a true Mexican system based on British, French, and German medical theories and techniques rather than the continuation of traditional humoralism found in Spain.[44] With this shift came an increase in responsibility and prestige for Mexican physicians and public health experts, who Mexican state officials believed held the key to improving public health.[45]

Similarly, historian Claudia Agostoni has argued that public health in the capital, Mexico City, improved during the Porfiriato due to infrastructural changes made by state officials. She argues that the desire by state officials to alter the physical shape of the city through the introduction of wider boulevards, parks, electricity, and the construction of monuments was integral to understanding how Porfirian officials believed they could create a so-called modern city. Agostoni refers to this process as symbolic legitimation, which involves the creation of visible and palpable representations of how officials defined modernity.[46]

At the same time, state officials also began to rely on physicians to serve as cultural arbiters of the progress underway in the city. This new recognition afforded them an opportunity to promote themselves in ways that reflected their newfound authority. Physicians began publishing thank-you letters patients had sent to them in local newspapers, which, Agostoni has argued, was how the medical community sought to strengthen its position in society and help contribute to the state's official narrative that the acceptance of science and technology would lead the individual down the path of modernization.[47]

Yet in all of these well-researched and informative works, one element that remained undiscussed was the dead bodies. While the dead "frolic vivaciously in the Mexican political and cultural imaginary," they also contaminated the living world of Porfirian Mexico City, where countless hundreds remained scattered, unburied, decomposing in city streets.[48] The guiding question that this book seeks to answer is how would state officials try to project an image of modernity in the face of the growing number of dead bodies? Moreover, how did city residents (upper, middle, and lower class) react to the various approaches employed by state officials to combat this corpse problem? This work is one of the first to examine how technology and science were integral components of the Porfirian modernization process that struggled to remove dead bodies from the city environment. The goal of this process was for state officials to improve the health of the city, which in turn would help to improve its reputation and hopefully put it on the same level as major cities in western Europe.

My work also helps scholars understand the relationship between the state and its people with death revealing the links between policy and daily lives in the capital city. It further serves as a bridge between the existing literature on Bourbon-era public health reforms and the measures undertaken by the postrevolutionary governments to continue to create hygienic, healthy, and productive citizens.[49] It also supports the existing scholarship that has explored how everyday citizens sought to challenge the desires of state officials to control behavior in the name of modernization. In this way, my book reinforces what James C. Scott and Bruno Latour have presented about the process of modernization. The idea that state officials could control and separate nature and society into distinct categories has revealed that try as they might to achieve this goal, they have failed repeatedly. As a result, the attempt by Porfirian state officials to capture modernity or any one of its synonyms, such as order and progress, should be viewed as the metaphoric dog that chases its own tail.

Porfirian state officials employed medical science and technology in matters related to death as tools for developing a modern society. In particular, they used new forms of transportation for corpses, promoted the utility of dissection in medical schools, and adopted new funerary technologies, all of which they believed would alleviate the city's corpse problem and improve public health in the capital. Yet the majority of the population, the urban poor, failed to adopt any of these changes. Instead, they chose to approach death in ways that made sense in their lives, which challenged the vision Porfirio Díaz had for Mexico.

Porfirian Population Management: Constructing Citizenship in a Modern City

Mexico's intense program of modernization, characterized by massive urbanization and rapidly changing demographics, took place during the thirty-four reign of Porfirio Díaz (1876–1910).[50] According to historian Alan Knight, "the Porfirian regime gave Mexico a generation of unprecedented peace and stability," which contrasted with the endemic political conflict the country experienced after gaining independence from Spain in 1821. Yet this Pax Porfiriana, as historians have named the era, was, of course, a flawed peace, based on the repressive political techniques of Díaz and his state officials.[51] Critical to understanding how state officials sought to transform this Pax Porfiriana was the introduction of population management techniques that attempted to judicially apply reason science, and technology to the lives of everyday citizens. This book examines how Porfirian state officials created a discourse that linked death, public health, medical science, and technology into a cohesive narrative that attempted to establish a vision of what being modern should look like in Mexico City and, they hoped, the rest of the country.

For the Porfirian government, as for the nineteenth-century government of the United States, the ideal citizen was passive and submissive and, ultimately, accepted that the state operated "in a benevolent manner."[52] While the dead do not appear to play an active role in citizenship, they do play an important role in the national imagination by suggesting an existence, both posthumous and posthistorical, that falls outside standard views of politics.[53] Dead bodies became "useful and effective symbols for revising the past," a useful method for Díaz to claim political legitimacy over the nation.[54] For the modern state, dead bodies were not meaningful in themselves, but when state officials used them to create a relationship that tied death, the state, and citizens together, dead bodies became "useful tools for others to put words into their mouths—often quite ambiguous words."[55] In Mexico City in the late nineteenth and early twentieth century, the dead spoke loudly about the problems surrounding public health in the capital and about how science and technology presented Mexico with an opportunity to eliminate the persistent stench of death on city streets and inside homes.

However, offering universal definitions of *modernity* and *modernization* is a difficult task. The set of characteristics used to identify either word changes drastically based on the audience, region, and period in discussion. Historian Paul Forman, nevertheless, has identified four cultural values that he has deemed necessary for understanding how best to define *modernity*: procedularism, disinterestedness, autonomy, and solidarity. These four values, along with an emphasis on the notion of self-discipline, Forman argues, "are peculiar to modernity." Furthermore, he has found that during the nineteenth century there was an increasing use of the terms *profession*, *professionalism*, and *professionalization* among scientists and physicians as a way to distinguish their occupations from the more traditional term *discipline*. The rise of these P-words led to changing definitions that centered on the concept of self-discipline. In particular, these definitions sought to create knowledge production, knowledge curation, and knowledge application enterprises that, as Forman points out, focused on the importance of autonomy, which had the "hallmark of modernity stamped all over it."[56]

This is the setting in which we find public health, medical education, dead bodies, and technology intertwined with the desires of the Porfirian state in Mexico City to control the health of citizens and how they mourned, celebrated, and interacted with death. This was part of a global modernist impulse that emerged in the late nineteenth century to equate improved public health and the hygienic practices of living residents with good citizenship and thus national progress.[57] Consequently, as historian Nayan Shah has argued, what we must keep in mind regarding these changes is the fact that "modernity, on one hand, promotes universality and, on the other hand, obsessively objectifies difference."[58] In the case of Mexico, this was part of a larger movement throughout Latin America that attempted to erase the traces of a country's indigenous past, which state officials considered barbarous and detrimental to achieving progress. Instead, officials wanted to replace their indigenous history with a narrative that saw the Indian as a quaint figure of the past, as a figure who fit the mold of the modern Mexican.

State officials in late nineteenth- and early twentieth-century Mexico attempted to cultivate a modern image of Mexico City by employing science, medicine, and technology in matters related to death. Chapter 1 explores how state officials sought to modernize Mexico City by adopting new transportation methods for the dead. The introduction of specific rules regarding hygienic corpse transportation on board trains first appeared in 1887, which spurred the growth of new transportation techniques for moving corpses in and out of the city to protect residents from potential health risks. Nevertheless, tension emerged between state officials and lower-class citizens over how to use this technology and navigate the modern city in the exact way state officials desired. I argue that the transportation methods (railroad, modern carriage, and electric tram) implemented by Porfirian state officials were part of the Porfirian government's modernization process that sought to introduce *capitalinos* (Mexico City residents) to the benefits they would enjoy as citizens living in a modern capital, even if the process would kill them.

Chapter 2 examines how medical education became intertwined with the official Porfirian discourse on modernity. Improving public health in Mexico City required more than just changing how residents transported dead bodies. Understanding the intricacies of the human body, achieved by increasing the number of dissections performed at the National School of Medicine, would yield respect, compassion, professionalism, and above all else a trust between physicians and patients. With the help of President Díaz's personal physician, Eduardo Licéaga, the National School of Medicine worked closely with state officials to improve the health of citizens by, as I argue, encouraging them to embrace medicine and the advice of medical professionals in the name of modernity. As a result, this relationship relied on residents' heeding the advice of doctors to improve their health. Poor health in the city meant Mexico's reputation would continue to suffer. Yet if capitalinos' health could improve, the future was bright for Mexico.

Protecting the health of citizens was a vital component of the campaign to modernize the country. An integral element of this was the state's desire to improve public health through the introduction of new technology that focused exclusively on protecting citizens from the threats posed by decomposing bodies. Chapter 3 examines new funerary technology, as state officials referred to it, that allowed medical professionals to slow down the decomposition process in an effort to preserve (ironically) the corpse. This new technology also helped to extend the same class divisions that existed in the world of the living to the world of the dead. The families of the well-to-do were the only ones who could afford this new technology, which included fashionable coffins, hermetic coffins, burial vaults, and topical and arterial embalming. By 1908, however, state officials introduced a new technology that the lower classes could all afford: the crematory oven. This technology allowed loved ones to be disposed of cheaply, often for free, as well as achieving the state's larger goal of erasing the presence of the poor as their bodies became ash in a matter of hours.

Chapter 4 explores how the urban poor challenged the government's multilayered approach to public health improvements in the city. The Porfirian modernization project hinged on the control state officials could achieve when it came to the behavior and customs of lower-class citizens, especially in matters related to death. Yet many lower-class citizens challenged this approach to regulating their lives and bodies, choosing instead to approach living and dying in ways that made sense to them, which were in stark contrast to the modern vision state officials had for the capital.

Creating the Narrative: A Note on Sources

Despite the commercial success of television psychics such as John Edward, James Van Praagh, George Anderson, or Theresa

Caputo the dead cannot speak to us.[59] But this does not prevent historians from exploring how historical actors felt about the dead. In particular, the records of Porfirian state officials are important for historians because of the dearth of information they left regarding public health. They recorded their daily interactions, which have given us a better understanding of how they treated dead bodies and fostered relationships with medical professionals to create a definition of modernity that fit their modernization agenda. Improving society through hygiene and public health has inspired scholars to write tremendous historical monographs that have illustrated how important dead bodies were to government agendas of modernizing cities and even entire countries.[60] But what makes Mexico's public health campaign unique is that unlike campaigns in the United States, France, England, Sweden, China, or Japan, Mexico had to deal with unprecedented migration rates to its capital, Mexico City. Estimates from the mid-nineteenth century placed Mexico City's total population around 200,000, but by 1895 it had grown to 329,774 inhabitants, and by 1921 to 615,327, numbers that did not include the thousands of citizens who commuted into the capital on a daily basis from the sprawling suburban towns surrounding Mexico City.[61]

This migration included many individuals from both indigenous and mestizo backgrounds (Spanish and indigenous heritage), which meant they brought their own customs and traditions to an environment that state officials wanted to exert control over. Integral to this control was the condition of public health, which historian Heather McCrea has argued allowed state officials to construct a relationship between nation-building and modernity.[62] State officials sought to introduce population management techniques that reinforced the government's increasing autonomy. Thus, state officials obsessed with public health filled countless volumes of material in the Mexican archives, which has preserved in reality only a sample of their obsession. As it turns out, while the dead cannot speak, they can be spoken of, and this provides scholars with a unique window into the minds of how state officials, state institutions, medical professionals, and upper-, middle-, and lower-class residents felt about public health and the problems associated with modernization.

To understand how a diverse group of historical actors has approached the issue of public health and death, I have based my study on a variety of sources. State regulations and decrees, institutional reports, letters to the governor of the Federal District and president of Mexico, business contracts, correspondence reports from cemetery administrators and medical school professors, medical school curricula, and patent applications are just some of the sources that shed light on the goals, accomplishments, and pitfalls associated with the population management techniques employed by Porfirian state officials in the name of modernity. The Historic Archive of the Ministry of Health (Archivo Histórico de la Secretaría de Salubridad y Asistencia) includes a collection of laws passed during the Porfiriato regarding the transportation of the dead that illustrates how important control over dead bodies had become for Mexican state officials. The Historic Archive of the City of Mexico (Archivo Histórico de la Ciudad de México) contains cemetery records for Mexico City and surrounding towns, which provides information on the day-to-day operations of these cemeteries, as well as important events or problems that cemetery administrators encountered. Included in these records are detailed reports that outline employee behavior and the complex relationship between the cemetery, the citizen, and the Porfirian state, as well a variety of state institutions. The records of the Archivo General de la Nación (National Archives) contain collections pertaining to patent applications, criminal cases that involved dead bodies, including international incidents such as the death of American railroad worker William Scott. The Historic Archive of the National Autonomous University of Mexico (Archivo Histórico de la Universidad Autónoma de México) maintains important collections related to the National School of Medicine, such as personal letters from professors, course catalogs and descriptions, and faculty, staff, and student records, all of which offer a window into how the medical school and state officials worked together to train physicians in an attempt to manage the health of citizens.

Constructing a modern image for Mexico was a complex process that reveals how state officials sought to incorporate public health, science, medicine, and technology in an attempt to manage the lives of citizens. However, this approach is never one-sided. The everyday actions of lower-class citizens demonstrate that the idea that state officials could control the behaviors and customs surrounding death or how individuals would use new technologies was an illusion. Their modernity would vanish as quickly as they could create it.

1

Moving into the Modern Era

Transporting the Dead in Mexico City

In the winter of 1904, a British visitor to Mexico City recorded her thoughts about how modernization was unfolding before her eyes. To her, there was no other city in Mexico where "modernity and barbarity shoulder one another more closely" than in the capital. On any given night, visitors and residents could lose themselves in the city's rich history, striking architecture, and cacophony of urban sounds. Amid the hum of electric lights, the thunderous clap of arriving trains, and the cries of street vendors selling corncakes, nuts, or charcoal, it was difficult to find a quiet moment. At the same time, however, the newfound highlights of modern urban life met a dark reality, a growing number of unfortunate souls who peddled and begged visitors for money to afford the costs of living in a rapidly urbanizing city.[1]

As a result, many of these individuals died on the street or in poor houses, a fact that was conspicuously missing from the British visitor's description of the capital. In fact, the number of corpses of the urban poor found scattered in city streets was overwhelming. These corpses were everywhere, including near the homes of the well-to-do, which threatened to ruin the progress that many of the city's elite believed President Porfirio Díaz and his state officials were steadily achieving in the capital.[2]

The majority of these dead bodies belonged to lower-class residents, many of whom had moved to the Federal District from other states throughout Mexico. The reason for such growth owed much to the development of the extensive network of railways that exploded during the late nineteenth century. For President Díaz and his cabinet, railroad development was essential for faster transportation of agricultural products and mineral resources, which promised "economic prosperity through export-oriented economic growth," political stability, and a new identity for a united Mexico free from isolated regions ruled by local *caudillos*.[3] The growth of the railways also introduced ordinary Mexicans to the marvels of modern technology that could transport them through time and space much faster than traditional methods of transportation could. For example, the completion of railroads linking Mexico City with the northern states of Chihuahua and Nuevo Leon in the 1880s contributed to significant population growth in the north, which peaked at 3 percent annually.[4] As historian Teresa Van Hoy has pointed out, the railroad made longer-distance travel take hours rather than days on foot or horseback.[5]

The growth of the railroad system was part of a larger economic development throughout Mexico begun by President Díaz. In addition to railroads, the government also promoted the expansion of public primary education, large-scale public works, and commercial agriculture. Moreover, Porfirian state officials also encouraged foreign investment in mining and manufacturing, all of which led to displacement among rural families who worked as agricultural laborers or artisans.[6] Beginning in the 1890s, Mexico City experienced unprecedented population growth. By 1900 migrants accounted for more than half of the capital's population, outnumbering residents born in the capital by two to one. Between 1877 and 1900 the population in the Federal District grew by two-thirds to more than half a million inhabitants. This trend continued throughout the early twentieth century: the population in 1910 was 720,753 compared to 327,512 in 1877, which represented a 120 percent growth rate in three decades.[7] The majority of new arrivals came to the capital looking for work. These workers earned their living through domestic work, commerce, delivery of services, production of consumer goods, and, after 1900, heavy industry, following the widespread electrification in factories.[8] What made Mexico City's working class unique compared to other Latin American cities was the fact that it included a high percentage of indigenous workers.[9] Many of these workers brought with them traditional customs, especially concerning daily life, of which hygiene and death were two essential components; both hygiene and death were integral to Porfirian state officials' recipe for progress.

Another ingredient essential for progress was new technologies, none more important than the railroad. The growth of Mexican railways in the 1880s made traveling around the country far easier, especially to Mexico City as well as within the capital, as the railways that converged there reshaped the capital's relationship to the countryside. For example, the Mexican Central Railway, the largest of the railways, allowed passengers to travel almost two thousand kilometers from Mexico City to El Paso at rates much faster than previous transportation modes. On the way to El Paso, trains also passed through an additional ten states, including Querétaro, Jalisco, Zacatecas, Durango, and Chihuahua. The route also included two lines to Guadalajara (Northwest) and Tampico (Northeast), which meant if the railway expanded from these points, it would ensure that both goods and people could move freely throughout the country.[10] Ticket prices varied in the Porfirian period, based on distance traveled. However, on average, railroad companies charged passengers between one and a half and three centavos per kilometer, depending on the type of ticket purchased (first, second, or third class).[11] Thus, the train allowed people of both meager and sizable incomes to travel great distances much faster than they could on foot or horseback. The constant influx of people, however, added to the already growing number of poor local residents, whom the government considered dangerous and dirty, impediments to progress.[12]

The tremendous population growth at the end of the nineteenth century created tremendous pressure on the capital's infrastructure, which led to poor residents living in squalor. Rising real estate values in the city center led to distinct residential patterns of class segregation in the capital. As workers constructed elegant mansions along the Paseo de la Reforma, the number of slums in the city ballooned around railroad stations, hospitals, public works facilities, slaughterhouses, and textile factories. In these neighborhoods, the city services provided in wealthy neighborhoods were absent, resulting in poor sanitation and overcrowding, which led to widespread health problems among residents.[13] The growth of this group—collectively known as the urban poor—helped contribute to Mexico City's international reputation as one of the most unsanitary cities in the world, highlighted by the alarming statistic of a mortality rate of 33.6 per 1,000 in 1900.[14] While major epidemics diminished as the nineteenth century progressed, largely due to improvements in sanitations and dedicated public health efforts by government officials, some records show life expectancy to reach as high as 40 years of age, while other data indicates a life expectancy of between 25 and 30 years of age.[15] Regardless, the fact is that the urban poor commonly resided in *vecindades*, one- or two-story tenement homes, where several families lived in single- or double-room apartments with doors that opened into narrow hallways, which were typically used for hanging clothes or as makeshift kitchens or bathrooms. With such cramped living spaces, as well as lack of understanding related to modern hygienic practices, it was no wonder that disease spread rampantly through these "troglodyte dwellings," creating a high rate of death in these homes.[16] Furthermore, tourist guidebooks pointed out that the lower classes bore the brunt of the poor drainage system, which was "thoroughly and radically bad, incorrect in its engineering, and ineffective in its results." Thus, diseases like typhoid as well as high mortality rates remained prevalent among the poor since many lived on the ground floors of buildings.[17] This was not the picture of modernity that state officials had envisioned.

To address the concern that both state officials and well-to-do residents had concerning public health and disease in the city, the government created a series of health and legal measures that sought to maintain invisible boundaries in the city separating the *gente decente* (elites and middle classes) from the diseased (the urban poor). Influential officials, writers, lawyers, doctors, and engineers known as *científicos* offered their *expertise* in order to remedy the situation. Influenced by Comtean positivism, científicos believed that in order to solve national problems, the government needed to apply the scientific method to all situations, especially those related to public health. They argued that all parties involved in presenting potential solutions had to agree that society was a developing organism and not a collection of individuals. Thus, the key to developing a stronger society (and government) was the introduction of scientifically driven laws and policies that would diagnose and solve problems quickly. But for an older generation of state officials, this approach conflicted with classic liberalism, especially as it concerned constitutional law and individual rights.[18] Nevertheless, científicos were successful at introducing policies that created an opportunity for the state to quantify people. Births, deaths, and housing were all integral components of Porfirian officials' method to identify and solve the city's hygienic problems.

As a result, statistical information intertwined with experts' assumptions about the exact causes of death and disease among the urban poor. Accordingly, the poor were the group most likely to become criminals, drunkards, and prostitutes, which exposed them to disease and an early death. State officials, however, failed to realize that as urban growth occurred, the lives of both elite and poor citizens intersected on a daily basis.[19] Elite neighborhoods required labor, labor provided by members of the lower classes. Lower-class women often worked as seamstresses or as domestics in the homes of the elite, while lower-class men worked as unskilled laborers, clearing trees to build roads and hauling materials required for constructing homes, among a variety of other labor-intensive jobs. Thus, workers who spent the majority of their days in these elite neighborhoods brought with them to the job their food, drink, and social customs, which often clashed with those of the city's elite residents.[20]

Furthermore, tension emerged between the urban poor and the upper classes over how lower-class residents should use public space, especially when that space crossed class lines. The government had attempted to create invisible, artificial boundaries throughout the city in an effort to protect upper- and middle-class citizens from the urban poor. However, when it came to the disposal of corpses, the Porfirian state constructed corpse deposits in areas of the city that did not belong exclusively to the urban poor. Rather, deposits existed in parts of the city where class overlapped because corpse removal methods (horse, electric tram) could reach them there more easily and quickly, which reduced exposure to potential public health problems found in decomposing corpses. Immediately, residents filled deposits full of corpses to the point that corpses began to spill into the streets in front of the deposits faster than officials could remove them, which created problems for the modern image of the capital that state officials sought to maintain. Moreover, these scenes also threatened upper- and middle-class values, especially those related to death, hygiene, and health, since these residents often walked by or near deposits in order to get to work or visit friends.

The introduction of deposits coincided with a decades-long push by the modern liberal state to transfer burial responsibility from the Catholic Church to the secular government. Burials were an important part of the modern state's population management techniques. Exerting control over the bodies of citizens (living or dead) became a hallmark of late nineteenth-century biopower around the world as the role of the state changed from "that of a saver of souls to a governor of bodies."[21] This governorship also provided economic opportunities since burial fees had traditionally provided the Catholic Church with significant economic earnings. For example, in the eighteenth century, a burial underneath the church and near the Eucharist cost twenty pesos, a fee only upper- and middle-class parishioners could afford.[22] These burial fees became an important portion of a priest's income, which would soon line the pockets of the modern state.

Rather than individual priests being responsible for burials, thus ensuring economic success for opportunistic individuals, the state homogenized the process, making itself the beneficiary in the process. Burial fees remained high, even when the modern liberal state put itself in charge of the burial process after Mexican independence.[23] The poor could not afford the rising costs associated with burials, and by the late nineteenth century, Porfirian officials seeking to avoid the potential pitfall of endemic disease and showcase benevolence offered free burials to residents who chose to leave their dead at designated deposits in Mexico City. As a result, hundreds of corpses appeared at the deposits, guaranteeing free burials at a rate that had been unimaginable to state officials.

The threat corpses posed to public health forced Mexican state officials to find potential solutions that would help reflect their desired image of the capital. This chapter explores how state officials tried to accomplish such a feat through the adoption of new transportation methods for the dead that moved corpses from city deposits and streets to public cemeteries. In particular, to combat the spread of disease, officials introduced specific rules regarding hygienic corpse transportation on board trains in 1887. This paved the way for additional methods for moving dead bodies safely throughout the capital, which other cities could adopt in turn as well. To state officials, these techniques would protect residents from exposure to potential health risks. At the same time, however, the urban poor reacted to these new methods in ways that challenged the government's idea that their lives and bodies belonged to the state. I argue that the transportation methods (railroad, modern carriage, and electric tram) implemented by Porfirian state officials focused on introducing capitalinos to the benefits of modern living and technology. Moreover, these same officials sought to strip citizens of essential rights, including self-ownership, in order for the state to protect the population as a whole.[24]

The Hygienic Railroad

Integral to the development of modern Mexico, including the capital, was the growth of the railway network in the late nineteenth century across the country. The expansion of the railway was part of the Díaz administration's official discourse that created a new nation that was nothing like the past, "a past characterized by economic stagnation, internal divisions, and constant threats from abroad."[25] The use of the railroad for transporting corpses throughout Mexico became one of several symbols of modernity for President Porfirio Díaz and his administration that demonstrated their commitment to progress.[26] By 1877, President Porfirio Díaz's first year in office, the country only had 640 kilometers of railroad track, but less than a decade later, by Díaz's long second term as president, the country had built almost nine times as much track (5,731 kilometers), creating an extensive railroad network throughout the country with Mexico City at the center. This was just the beginning. At the end of the nineteenth century, Mexico's track measured 12,173 kilometers, and by the time Díaz left the presidential office in May 1911, it had grown to almost 20,000 kilometers.[27] The expansion of the railroad became synonymous with Díaz's presidency, and newspaper writers, both liberal and conservative, praised the president for his effort to develop Mexico.[28] The railroad became a way to unite the country, bringing civilization to both urban and rural Mexico while also providing all with a glimpse of what the future held.

Despite the excitement and celebration surrounding the railroad, problems remained. In addition to the displacement of indigenous groups, especially the Yaquis and Mayas, there was little concern about public health on the railroad.[29] In fact, the railroad industry continued to take few precautions when transporting corpses. Bodies traveled freely between Mexican states and even into the United States, with no specifications governing the types of materials used to secure the body; the only requirement was that a signature appear on a certification of embalmment. Thus, it was possible that passengers could travel in or near the same cars as the dead.[30]

By 1887, however, during Díaz's second term, his administration introduced specific rules that emphasized to railroad employees the importance of hygienic protocol when transporting corpses.

The new rules covered issues such as the construction and packing materials that surrounded corpses in coffins as well as the inclusion of demographic information for the deceased. Once the body reached its final destination, the new regulations required that workers follow specific disinfection procedures once they had unloaded corpses from cars.[31]

Segregated physical space was another way Porfirian state officials ensured the health and safety of those onboard. Corpses no longer rode in the same cars as passengers. Instead, workers put coffins in the car at the back of the train, as far away as possible from passengers and crew. Once the train had delivered all of the corpses to their final destination and passengers had disembarked, railroad employees cleaned the cars that had contained corpses with an antiseptic made from carbolic acid or Labarraque chloride (an early form of bleach). This solution eliminated the presence of any bacteria or odor associated with the dead.[32] Transporting bodies by rail could be dangerous to workers and passengers, and the explicit rules instituted by state officials reveal how important maintaining hygienic conditions was to them.

The new regulations and materials governing the construction and packing of coffins reinforced the importance of public health. More importantly, these changes helped to create a set of modern transportation standards that sought to shape the image of Mexico and demonstrate the government's dedication to modernization. The rules now required workers to place corpses transported by train into two coffins, each made from zinc, lead, or galvanized iron and with walls at least three millimeters thick. Like a set of Russian nesting dolls, the smaller coffin fit inside the larger one. A preservative powder made from sawdust and zinc sulfate, or carbon dust and tree bark, was to surround the corpse inside the smaller coffin and fill any empty space that existed between the two coffins. Once fastened by screws or nails, the regulations stated, coffins became hermetically sealed.[33] State officials believed that these changes would prevent noxious gases and odors associated with decomposition from escaping coffins and possibly endangering the lives of passengers and crewmembers onboard as well as residents living near the railway.

The second of the government's new rules addressed the origins and demographic information of the deceased. A form would now accompany the coffin, listing the deceased's name, age, and date of death, together with a signature from a doctor or "competent person" who had verified these data. The form also contained a separate section to fill out if the individual had died from a particular disease. If the cause of death were typhus, typhoid fever, diphtheria, smallpox, or Asiatic cholera, common diseases in Mexico during this era, than standard transportation of corpses was possible. In particular, Mexico City and central Mexico were susceptible to smallpox, cholera, and typhus epidemics largely due to their dense populations, which made the transmission of disease favorable; overcrowding and poor sanitation exacerbated the situation.[34] However, if the cause of death were yellow fever—a disease that came from mosquito bites and was more prevalent in environments with tropical temperatures (between 90 and 102 degrees Fahrenheit) and substantial humid conditions like those found in Veracruz or Chiapas—the government restricted transportation to locations above sea level or regions where the climate was not conducive to the spread of the disease.[35] State officials required similar documentation for Mexican citizens who died outside of Mexico or for foreigners who requested burial in Mexico. Finally, the state required an adult escort to accompany the corpse to vouch for the validity of the demographic information.[36] The government's desire to keep detailed records to identify the cause of death was an important part of the Porfirian attempt to regulate health (and death) as part of an ongoing policing project that sought to showcase the state's larger investment in its population.[37]

Railways and Corpse Deposits

Hygienic rules concerning the transportation of corpses on trains helped shape individual perceptions about how the railroad would become a useful tool for dealing with the capital's growing corpse problem. For state officials, adherents of positivist ideology, Mexican society was an organism that used its network of railways as its arteries, "pumping blood and life into a moribund nation."[38] Thus, the railroad presented officials with an opportunity to showcase how new technology and life should coexist harmoniously in the modern world.

Porfirian officials had failed to prepare for the unexpected as Plaza Concepción became a popular location for the urban poor to dispose of corpses. Soon, bodies began to spill into the street, where they remained a threat to public health. Located a few blocks northeast of Alameda Park and Hospital San Andrés, this threat provided state and city officials with the perfect opportunity to illustrate how best to combat hygienic problems. The choice of Concepción as a location for corpses coincided with the existence of a nearby defunct colonial era cemetery, Santa Paula. In March 1889, rather than allow the growing number of dead bodies to continue piling up around Plaza Concepción, state officials decided to build new railroad tracks that would connect the deposit to Santa Paula Cemetery. This, officials believed, would reduce the potential threat decomposing corpses posed to the health of city residents. Removing corpses by railroad presented state officials with a unique method for ameliorating public health challenges in the city. Furthermore, the creation of another railroad, an invaluable symbol of progress in the Porfirian period, demonstrated the state's growing authority over death and reminded residents how inextricably tied the railroad was to modernity.

Panteón Santa Paula dated back to 1836, when Archbishop Don Alonso Nuñez de Haro y Peralta established it as a location for the burial of the poor who had died at Hospital San Andrés, one of the church hospitals in the city. In the beginning, the cemetery was a simple field (measuring 225.7 meters by 117.9 meters) surrounded by a wall with a small chapel located in the middle, but it grew to include 1,665 grave sites over the next two decades, until an earthquake almost destroyed the entire cemetery in 1858. Nevertheless, the cemetery remains intriguing because of the role it played during the cholera epidemic in 1850 (May 17 to August 2) that killed more than 5,000 people in Mexico City. The cemetery also became more ornate over the years, as graves covered with metal or marble with epitaphs and poetry written in polished gold or silver became more common.[39]

Despite the improved quality surrounding the graves at Santa Paula, many city residents still considered it to be a danger to public health. Various seasonal changes in the city, especially with a shift in winds, residents argued, could mix with gases emitted by decomposing corpses. With so many corpses buried in the cemetery, many of which workers had covered using a mixture of lime and coal dust, led to residents seeing Santa Paula as a site that "allowed miasmas to freely escape into the air and bathe the capital."[40] The government closed the cemetery permanently in 1871, and following its closure it was left unkempt, becoming a site of "repugnant desolation, distress, and repulsion."[41] According to several influential Mexican writers during this period, there were two fundamental problems with cemeteries: poor burial planning by the Catholic Church and the backward behavior of cemetery workers and lower-class visitors. Famed Mexican writer Ignacio Manuel Altamirano lamented the fact that the Catholic Church had failed to provide sufficient burial space in cemeteries like Santa Paula, which led to the creation of "gloomy, horrific, and anti-hygienic spaces."[42] More writers like Antonio García Cubas, Manuel Rivera Cambios, and Marco Arróniz also pointed out that the people of Mexico required retraining when it came to how to use cemetery spaces, a message that resonated with Porfirian state officials. Events such as Día de los Muertos (Day of the Dead) resulted in uncouth behavior as individuals "enjoyed themselves too much, laughing, wearing their best clothes, eating and getting drunk every Second of November."[43] Some writers even lamented how cemeteries like Santa Paula, once beautiful and landscaped spaces, had become nauseating environments where workers placed bodies of the poor in shallow graves, sprinkled them with lime or carbon, covered them with a little bit of dirt, and laid loose planks of woods over the hole.[44]

As Manuel Altamirano put it, as the population of Mexico City grew, the city would become "a volcano of pestilence, an exterminating angel, and a perpetual threat." To improve this situation, Altamirano and others urged cemetery administrators and state officials to plant more trees and flowers in order to "ventilate the space and dissipate gases" coming from graves. These changes were only necessary in Mexican cemeteries in the city and not in the French, English, or American cemeteries, where there was a "slight breeze" with lots of trees and flowers and corpses buried properly beneath the soil, which provided a far different perspective and condition for death than how Mexicans had treated their dead in the nineteenth century.[45] The capital required new cemeteries because those that already existed both were unhygienic and lacked the necessary burial space to keep up with the growing number of people arriving in the capital daily.

Such disgust would not disappear during the next two decades as the Porfirian government reopened the defunct cemetery to solve another public health problem. As state officials expedited the plan to connect the railroad to Santa Paula to facilitate the burial of corpses from Concepción, they had failed to investigate whether Santa Paula was private or public land. The company responsible for building the railroad, La Compania Limitada del Ferrocarrilles del Distrito Federal, informed Federal District secretary, F. L. del Castillo, that Santa Paula was not public land. Instead, it belonged to Don Ygnacio M. Escudero, a citizen who prohibited the company from installing railways tracks inside the cemetery.[46] Thus, the construction of the railway's tracks stopped just short of the cemetery gates. Construction on the railway intended to link the deposit at Plaza Concepción to Santa Paula Cemetery ended almost as soon as it had begun.

Nevertheless, city employees continued to deliver corpses from Concepción to Santa Paula. The number of corpses found outside the deposit at Plaza Concepción had decreased, but the public health nightmare had just shifted locations. Now corpses lay piled in front of the gates at Santa Paula. Unable to bury the bodies, and with seemingly no other alternative, corpses remained stacked in piles like sandbags.[47] The result was an unflattering image of the supposedly modernizing capital—one that would continue to exist for five more years. Putting the best face on the situation, the government maintained that despite the setback, the railroad had improved public health for people near Plaza Concepción.

Yet not everyone agreed. While the government believed the railway offered city residents tangible proof of modernity by protecting public health through the removal and transportation of bodies, Eduardo del Valle, the president of the railway construction company, disagreed. He wrote a letter to Federal District authorities to express his outrage over the administration's continued use of the railway to transport corpses to Santa Paula despite not being able to bury them. The result, del Valle argued, was disastrous for public health. As he put it, the government "should expedite this matter since it was the one responsible for splitting up and selling the land in question." But state officials ignored him. Instead, they insisted that the railway continue to operate as originally intended.[48] The city continued to deliver bodies from Concepción to Santa Paula because it not only improved public health for a more populated area of the city but more importantly served as an important instrument for legitimizing new ideal behavior and technologies.[49]

The Inconsistencies of Modernization

The situation at Santa Paula, however, illustrated the paradoxical nature of modernization in Mexico City. While many city residents and state officials applauded the transformation of the city, other residents did not. In fact, many of the so-called benefits of modernization came at their expense. In order to showcase a variety of dazzling modern symbols like large steel buildings, parks, street lights, or railways, state officials often targeted lower-class neighborhoods for demolition.[50] They also believed that these changes would lift working-class culture, traditions, and health to a level never before seen, in spite of the uncertainty surrounding basic needs of lower-class citizens, such as where they would live, eat, or sleep.

It is certainly the case that before the presidency of Porfirio Díaz, the Mexican economy had suffered from a lack of transportation and communication facilities, as well as limited banks, capital, and technology. During Díaz's tenure, the country did experience uninterrupted growth with booms in railroad construction, foreign economic investment, wealth, and population—all factors that Díaz and state officials believed were important contributions to the modernization process in Mexico.[51] Nevertheless, this growth was spotty and even contradictory at times, characteristic both of an underdeveloped economy and uneven modernization process.[52] For example, state officials were not particularly concerned with the social implications that their modern plans produced. Mexico City, along with other major Mexican cities at the time, experienced increasing economic disparity between social classes, low wages, high food costs, and rising crime rates, which served as a reminder to those affected (working-class residents), that state officials paid little attention to how these changes impacted their day-to-day lives. Mexico would become modern, even if it meant that all the poor residents had to die. In the capital especially the government believed that its new modern toys, such as trains, sidewalks, electric lights, and indoor plumbing, would bury any remaining hints of barbarism.[53]

From 1889 to 1894 the continuous delivery of corpses had exposed nearby residents to nauseating sights, unhealthy smells, and potential disease. On April 26, 1894, Governor Pedro Rincón Gallardo found a method to protect the future health of the city in a way that demonstrated the accuracy of equating modernity with the capital. In that year, he persuaded Don Ygnacio M. Escudero, the fifty-seven-year-old widowed owner of Santa Paula, to sell the cemetery to the Federal District for one thousand pesos ($14,500). Included in the sale was a tract of land adjacent to the cemetery that provided the government additional space for burying corpses.[54] Soon thereafter the railroad company succeeded in completing the tracks that would allow transport of corpses into the cemetery. For state officials, the ability to bury the bodies of the urban poor would not only improve public health but also reinforce how public health had become part of the political and economic landscape of the modern capital.

Regardless, the haphazard nature of the modernization process in Mexico City had appalled hundreds of residents. Between July 1889 and February 1894—when the corpses of the urban poor remained in gruesome piles outside Santa Paula—citizens from surrounding middle-class neighborhoods forcefully expressed their displeasure with how President Díaz and his state officials delivered modernity to the capital. In letters to various governors of the Federal District, residents argued that their health and the health of the neighborhood was under constant threat from the corpses left decomposing outside the cemetery gates. The persistent visual and olfactory reminders helped to foster residents' distaste for how state officials framed the official narrative surrounding modernization.

This situation was nothing new for the cemeteries of Mexico City. Physicians in the 1870s and 1880s had written that the capital, more than any other urban area in Mexico, had suffered from "a plague of cemeteries" that had failed to follow hygienic protocol, thus producing an environment that was "a warehouse of putrefying human flesh."[55] To improve the situation, physicians argued that the state needed to consult with hygienists to help determine the proper size plot required for burying corpses, study wind direction in cemeteries, and record the level of decomposition of corpses, all of which reinforced the beginnings of a push for scientific expertise to be included in the debate over how to improve public health. But the changes needed to achieve the quality of public health that hygienists and physicians desired would face an uphill battle. For example, one medical thesis from 1872 found that Santa Marta, Santa Paula, and Campo Florido cemeteries all had the same problem; specifically, workers buried too many corpses in one large grave, anywhere from twelve to eighteen bodies. Such a quantity of corpses packed into a small space had led to repeated incidents involving both "dogs and buzzards preying on the remains of the dead."[56] Situations like this would not disappear by the end of the nineteenth century, as Porfirian state officials would soon find out, even as they devoted more time, energy, and resources to improving public health.

One such resident, R. García, expressed his disappointment with how the government had handled the potential dangers of the situation. He had written the governor twice, each time asking him to suspend the train service in the neighborhood. But the governor ignored García, never issuing a response to his grievances.[57] The reason: the railroad continued to present state officials with an instrument of unmatched civilizing capabilities. It had become an important component of middle-class and elite discourses concerning self- and class definition, which allowed the two groups to declare themselves the "vanguards of modern life."[58] These groups believed that the railroad was one of the best tools available for shaping the urban landscape, especially the relationship between working-class culture, hygiene, and the disposal of dead bodies. The railroad had ushered in a new set of cultural values that were part of profound social and technological changes occurring in Mexican society.

In particular, the emphasis on public hygiene overlapped with the government's desire to dispel the harmful influences of urban living. Hygiene was part of a larger trend "to regulate the flow of people, goods, necessities, and wastes through the city in order to redress the negative effects of having to share the same space with many others."[59] Led by aristocratic reformers, the government eagerly adopted newfound social engineering tactics that called for the modification of lower-class culture and behavior. This therapeutic package, which included the railroad, sought to eliminate presumed lower-class vices like alcoholism, gambling, prostitution, and crime. Reformers assumed the lower classes would quickly adopt more refined behaviors and tastes in order to dedicate themselves to sobriety, hard work, and hygiene.[60] Soon these individuals would become modern citizens.

Yet residents did not universally share this optimism. For residents such as García, it was one thing for the state to be the driving force behind eliminating the backward behavior of the urban poor. It was quite another to have the bodies of the urban poor cross the invisible boundaries that separated middle- and upper-class residents from lower-class ones.[61] He voiced his concerns clearly: the pile of corpses outside the cemetery "infected airs that could cause serious damage to one's health" in addition to offending the sensibility of the residents.[62] Nor was García the only citizen willing to voice umbrage with how the government was addressing the situation at Santa Paula.

Hejo González and Carlos Puchón, residents from a nearby neighborhood, wrote the governor as well. As García had, they expressed irritation that the governor never responded to their complaints. While the governor ignored them, the number of decomposing bodies had increased, threatening the water supply at a nearby aqueduct. The overwhelming presence of bodies had contributed to an invasion by "mounds of flies" that consumed the flesh of the corpses and laid eggs inside them, all the while using the water as a breeding ground for their larvae. As González and Puchón put it, the unburied corpses had turned their neighborhood into "a hotbed of decay."[63] This potential health nightmare coincided with the shared belief held by both state officials and well-to-do residents that the lower classes were "great enemies of hygiene, culture, and progress" and impediments to Mexico's successful adoption of modernity.[64] As a group, the poor had the potential to infect the rest of society, which jeopardized the government's ability to promote a modern image of Mexico City. The decaying bodies at Santa Paula undermined the progress that many reformers had hoped to achieve in the capital.

But the state and its sanitation army were not the only ones to offer solutions to the public health nightmares that plagued the capital. Santa Paula's neighbors did not merely complain; they also offered remedies of their own. To protect the health of the neighborhood, González and Puchón urged the governor to order state officials to discontinue using the railway that connected Plaza Concepción to Santa Paula. The two believed that the government instead could move the corpses to the southwest part of the city, where they had found available land adjacent to Belén Prison. Such a move could be more expensive than the government originally planned. However, according to the duo, this ought not to be the overriding issue since "it will not cost half of what matters, which is the lives of the residents of the neighborhood of Santa María and the said chapel of Santa Paula."[65] Their plan was intriguing because a railway already existed near Belén that would allow the government to continue to use the railroad and thus demonstrate its commitment to hygiene as well as modernity.

The location of the proposed new repository revealed a unique feature of modernization in Mexico. If the government chose to move the corpses from Santa Paula to Belén Prison, state officials (by fixing this problem) could continue to create a well-ordered hierarchical society. One influential aspect of modernizing the country involved assessing the value social groups could offer the country. The prevailing ideology—based on a combination of positivism and social Darwinism—placed greater value on the lives of middle-class and elite citizens. State officials argued that unlike the poor, these two groups contributed in a positive manner to society and as a result were an integral part of ushering Mexico into a modern era.[66] Empirical "proof" that these groups were superior came from scientists (biologists, physiologists, hygienists, and criminologists) who obliged the state by helping to create a discourse that argued for the inherent inferiority of the lower classes, who were morally degenerative and thus required isolation from the *superior* parts of society.[67] The availability of land near the prison—an institution already filled with many members of the lower classes—would allow the government to maintain its grip on society by removing the potential public health threat from middle-class neighborhoods and put the threat squarely where it belonged: among the urban poor.

Not surprisingly, residents like González and Puchón firmly believed in their own inherent supremacy. Relocating the deposit to Belén meant moving the corpses to a less populated area of the city, where the dregs of society lived. In turn, this would solve

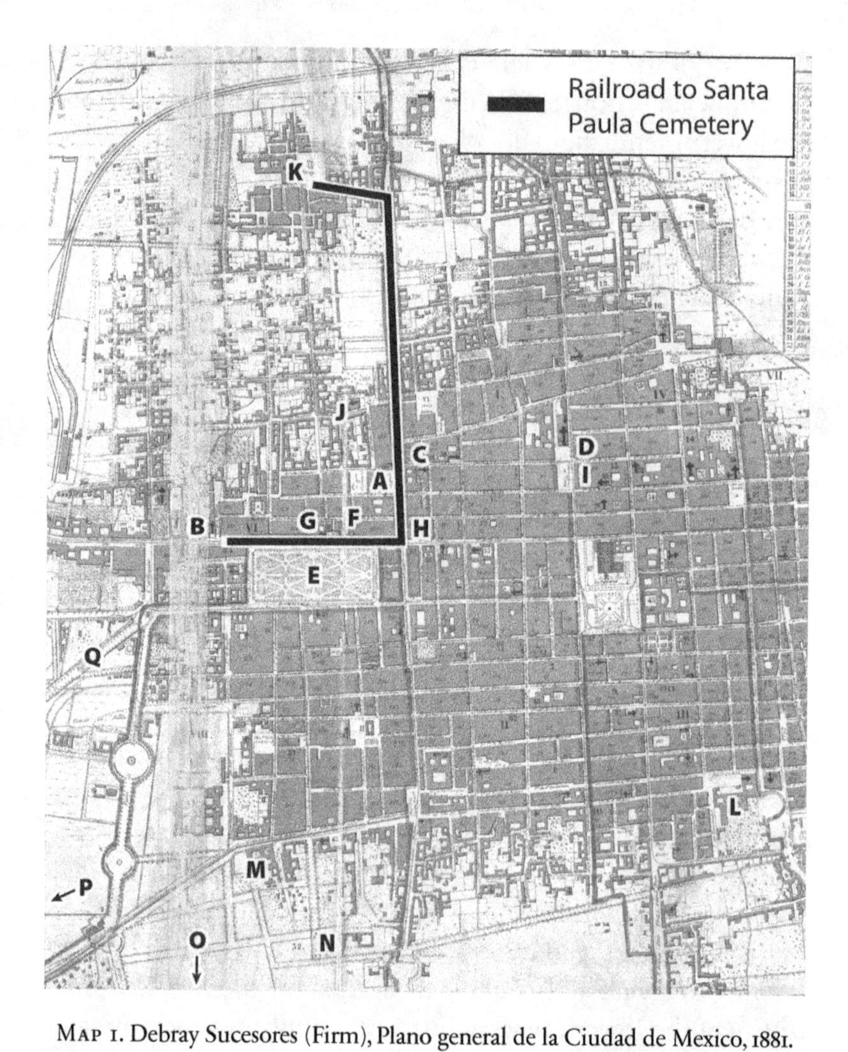

Map 1. Debray Sucesores (Firm), Plano general de la Ciudad de Mexico, 1881. The locations that most concerned state officials are marked C, F, J, and K, where the railroad linked the corpse deposit, local neighborhoods, and Santa Paula Cemetery. Map reproduction courtesy of the Norman B. Leventhal Map Center at the Boston Public Library. Key: **A.** Orrin Brothers Circus, **B.** Hospital San Hipólito, **C.** Plaza de la Concepción, **D.** Escuela Nacional de Medicina, **E.** La Alameda, **F.** Neighborhoods of Villamil/Carbonero, **G.** San Juan de Díos, **H.** Hospital San Andrés, **I.** Consejo Superior de la Salubridad, **J.** Neighborhoods of Santa María de la Redonda/Plaza de Jardín, **K.** Santa Paula Cemetery, **L.** Hospital Juárez, **M.** Belén Prison, **N.** Campo Florido Cemetery, **O.** Hospital General (approximation), **P.** Pantéon Dolores Cemetery (approximation), **Q.** Paseo de la Reforma, **R.** Colonia Cuauhtémoc.

their immediate concern: ameliorating the public health problems found in their neighborhood. Moreover, this would also create a more segregated city where safe distances existed between the civilized and uncivilized citizens, which would help paint Mexico City as modern. State officials considered Belén Prison to be a source of disease: hygiene was nonexistent there, and its poor location—combined with a lack of drainage—had turned the area into a virtual sewer.[68] As one tourist guidebook suggested, the prison was "a microbic spot that should be avoided."[69] The government deemed prisoners, many of whom were urban poor, to be a social group without any redeeming qualities, which made them expendable.[70] By moving the corpses to Belén, the government could improve hygiene in the Santa Paula neighborhood, placate angry residents, and continue to promote the idea that changes in the urban landscape were modernizing Mexico City before their very eyes.

Public Health and National Identity

Before acting on this intriguing plan, the government had to discuss the logistics with La Compania Limitada del Ferrocarrilles del Distrito Federal, the railroad company responsible for completing the railways in the city. While state officials believed González and Puchón's proposal would help reinforce the social order, improve public health, and elevate the city's modern image, Eduardo del Valle, the railroad company president, vehemently disagreed. He argued that in all the excitement about the potential change, the government had overlooked an important factor completing the project—the health of the railroad workers.[71] The area was, by all accounts, dirty and unhealthy. Del Valle bluntly informed the Governor Gallardo that his company would not install more track nor connect its railroad to existing railways near the prison.[72]

However, del Valle did present the governor with an alternate location that he thought would offer the same benefits but also protect his workers. He argued that the "most appropriate location" for the disposal of these corpses was in the colonial-

era cemetery Campo Florido.[73] There were three specific reasons, del Valle pointed out, why the government should use the defunct cemetery. First, the cemetery was in "the least populated area of the city" and would be "sheltered from the prevailing winds." Less wind meant better protection for workers and fewer chances that diseases would spread into more populated areas of the city, which was useful for preventing an epidemic from sweeping through the capital. Second, the government owned Campo Florido and the existing railway line just outside of it. To deliver the corpses by train, the government would only have to finance the construction of a "short stretch of railway." Furthermore, if the government moved corpses to Campo Florido, the railroad company could use the existing railway to create a line between Florido and Panteón Dolores, a popular public cemetery. This connection would provide the government with additional space for burying the corpses of the poor.[74] Del Valle's arguments shrewdly anticipated that state officials would place the utmost importance on maintaining hygienic railroad transportation for the dead. It is quite probable that the suggestion to use both cemeteries had to do with the fact that neither was the resting place for liberal political figures as well as the remote locations of each, which would safeguard health and demonstrate the Porfirian commitment to modernity above all else.

The Problem with Tangible Modernity

Yet despite the advantages of the Campo Florido proposal, Governor Gallardo rejected the move. Why the governor decided against the project remains a historical mystery: no records have come to light regarding why he turned it down. Nevertheless, complaints about the corpses remained. Citizens living near Plaza Concepción and Santa Paula continued to complain to the governor about the threat posed to them by decomposing corpses.

For several years, residents near the deposits at Plaza Concepción had also expressed their anger over the poor hygienic conditions surrounding their neighborhood. As early as 1890, two neighbors, Francisco Yglesias and Lucio Romero, had gathered

129 signatures from other area residents, which suggests that this was not isolated discontent by one or two curmudgeons. Writing to the governor, they expressed anger that the corpses of the urban poor constantly threatened their health. They pointed out that the constant presence of the dead could "easily compromise the rest of the population." Furthermore, the decaying bodies had attracted dozens of hungry stray dogs, which ate the decomposing flesh of the corpses, forcing the neighbors to pool their money to hire private security guards to stop the dogs from snacking on the bodies or occasionally dragging one through the streets. These horrors, shocking in themselves, also had commercial implications. According to Yglesias and Romero, their ability "to rent or lease properties in the area" had disappeared because nobody wanted to live or work in a neighborhood with such conditions.[75]

For several years, letters about the dead bodies found near Plaza Concepción continued to reach the governor. One of the more striking ones appeared in February 1894, when neighbors Vidal Diez and Mauricio R. Ramírez—along with at least thirty others—wrote to point out that the problems plaguing public health had affected the neighborhood and its businesses for far too long. The group wrote Gallardo to tell him that as long as the railway linking Concepción to Santa Paula was in use, "the number of germs" would increase exponentially. Additionally, they warned the governor that the winds in the area had the potential "to carry the evil" beyond the neighborhood and infect the rest of the city. But these two neighbors, like others before them, had a solution. If the governor moved the corpses away from Santa Paula "to somewhere in the southern part of the town," where fewer people lived, as the Campo Florido project had suggested, public health would improve.[76] Again, spatial segregation appeared to be the key to placating the fears held by middle- and upper-class residents. The situation at Concepción also illustrated the uneven process of modernization in Mexico City. While state officials had invested in corpse deposits to improve hygiene and public health, they had unintentionally exacerbated both and failed to demonstrate that the city was any closer to becoming modern.

Decomposing bodies remained a problem in Mexico City throughout the 1890s. The government continued to ignore the numerous complaints. Perhaps state officials were unable to find any empirical evidence to suggest that public health truly was at risk. But more likely, the key reason the government disregarded the grievances was that addressing them would have meant admitting defeat. President Porfirio Díaz and his officials, which included state governors, played a crucial role in creating, reinventing, and managing a national mythology that involved significant efforts to control and regulate the lives of citizens—including their dead bodies.[77] Perhaps addressing the concerns of residents meant state officials would have to acknowledge that they were not as good at managing and controlling society as they believed. If the state considered public health to be an investment—consider the effort officials put into defending and supporting new changes designed to maximize the value of citizens' lives—then admitting that citizens were at risk meant the state had made a bad investment.[78]

Additionally, the railroad was an important symbol of modernity that state officials wanted all capitalinos to share. Installing modern technology in populous areas of the city was strategic and outweighed any concern over health expressed by a minority of citizens.[79] Had the government moved the railway as residents had requested, the state's symbol of modernity would be less visible, and the impact of the railway as a tangible example of modernization would disappear. While continuing to use the railway between Santa Paula and Plaza Concepción, where thousands of people lived, the government could expose residents to the power of the Porfirian state. The railroad had become an integral pedagogical tool of the state, intended to teach the urban poor about the benefits of modernity.[80]

While the position of governor of the Federal District changed frequently in the 1880s and 1890s, citizens continued to express their concerns for their health and the health of the city. In June

1898, Governor Rafael Rebollar issued a press announcement that lauded the success that his cabinet had experienced recently concerning the removal of an overwhelming number of corpses found in the city.[81] Campo Florido now connected to Panteón Dolores via railroad, which should have meant improvements in public health. However, discontent among city residents continued to fester.

In areas outside Santa Paula and Plaza Concepción, locals had expressed their dissatisfaction and disgust with the stench and potential health hazard found at nearby corpse deposits.[82] While some problems had found solutions, another threat had emerged a few years later in an area of the city where the elite resided. In June 1903 Pedro Sobrino, a resident of a well-to-do neighborhood just off Calle de Arcos de Belén, wrote to the Superior Sanitation Council (SSC), the government institution responsible for public health and sanitation in Mexico City, to express his abhorrence of the "unpleasant and dangerous scene for one's health" he encountered on a daily basis outside his home.[83] According to Sobrino, the urban poor routinely filled a nearby corpse deposit to its maximum capacity. Once there was no more space remaining inside the deposit, instead of returning to their homes with the deceased, many of the urban poor chose to leave them scattered in the street: grotesque trail markers that led inhabitants to the deposit.[84] City workers ignored the problem, allowing the corpses to remain for hours (even days) in the street, where they threatened not only the health of surrounding neighborhoods but also the government's desire to maintain a modern image for the city.

Unlike previous complaints, Sobrino's would foster change in the collection of bodies. Expressing concern over a potential health epidemic that could develop from the situation, or most likely because of his socioeconomic background, Superior Sanitation Council members decided that it was "highly desirable to prevent the people of the town from laying out the bodies of their relatives on a public road."[85] The collection of the dead by city workers had previously occurred twice a day: from 6:00

to 6:30 in the morning and from 3:00 to 3:30 in the afternoon. Both times were inconvenient for many of the urban poor, who worked during these hours. Newly elected Federal District governor Guillermo de Landa y Escandón believed the existing schedule had failed "to prevent the bodies from being deposited, even for a short time, on the streets of Arcos de Belén." The corpse deposits had limited space and thus filled quickly. Deposits were first come, first served, which meant that those who were unable to get there early were left without an opportunity to dispose of the deceased properly, so corpses littered the street. Not known for radical changes, the Superior Sanitation Council officials first decided to combat the problem by continuing to use the method already in place, but with one change—they decided to have the car that collected corpses "stay in the neighborhood for an extended time."[86] In the opinion of the council members, this would allow more of the urban poor to drop off corpses at or closer to the deposit and guarantee better success in the collection of the deceased. By adjusting pickup schedules, the Superior Sanitation Council believed it would offer goodwill to residents, perhaps shape their behavior, and more importantly improve the quality of public health. All of this would help strengthen the official narrative of order and progress that positivist elites believed was essential to creating a modern Mexico.[87]

The governor also used his authority to suggest to other state officials that the city should finance the construction of an additional corpse deposit near the problem-plagued neighborhood Pedro Sobrino had described.[88] If altering the collection times for bodies did not remedy the situation, an additional deposit site would surely alleviate the corpse problem and thus protect public health and keep the streets clean. Whether the government constructed the additional deposit site remains unknown—further information has disappeared from the historical record. Nevertheless, the proposal is important because it provides two important clues about how the modernization process unfolded for various social groups in Mexico City. First, the urban poor knowingly ignored concerns about public health expressed by well-to-do residents, city institutions, and state officials. The changes to the schedule failed to make an impact on how lower-class citizens disposed of corpses, as they continued to leave corpses outside designated areas, thumbing their noses at these officially prescribed rules.[89] Second, both the schedule change and potential new construction overlapped with the project of modern statecraft. Public hygiene, which is situated at the "intersection of biological, moral, and political economies," was a fundamental part of this project. Couched in official rhetoric that described the process as a method for civilizing the population, the chief goal was to shape the population and landscape in a way that would "link the needs of society to the apparatuses of the state."[90] Thus, lower-class residents would become more susceptible to the persuasive techniques of state officials who wanted to observe and control their behavior. In the end, the poor would conform (*hopefully*) and thus allow the Mexican government to create a homogenous citizenry imbued with the moral and cultural values that reflected modernization.

This approach had found inspiration in mid-nineteenth-century Paris, where Georges-Eugène Haussmann had experienced success in modernizing Paris. The Paris that Haussmann improved had only recently expanded. French emperor Napoleon III had torn down the Farmers-General (Octroi) Wall, erected in 1791, which had surrounded the city in order to expropriate nearby valuable land. The city's landscape until 1850 had remained a relic of the past, a town that state officials and visitors considered "dark, dirty, foul-smelling, and overcrowded."[91] But Haussmann's renovations sought to open the "dark, confined, and frightful city" by demolishing significant parts of it—most notably several major streets and working-class neighborhoods. In their place, he created large boulevards (400 miles of new pavement that made the narrow and curvy streets longer, wider, and straighter), uniform façades for buildings (he tore down large, dilapidated apartment buildings and gabled homes that had been subdivided to accommodate more residents), public green space (47,000 acres versus 47 acres, and 100,000 newly planted trees), new sewer systems (260 miles of new lines), and a new aqueduct that brought fresh spring water from the countryside into the city, replacing the inefficient water collection method that had relied on thousands of water carriers to deliver daily fresh water to residents.

In addition to the hygienic advantages these features would provide state and city officials, they facilitated the state's ability to collect important information that pertained to the population. Redesigned city streets made mobilizing troops and police more efficient—a problem that had plagued Paris in earlier years—by placing army barracks at the intersections of major thoroughfares.[92] It also created distinct neighborhoods segregated by class and occupation, which permitted government officials "to easily manage and administer" the lives of Parisians.[93] These changes served as an inspiration to Francophile Mexican state officials, who would copy what Paris had done, tweaking the changes slightly to fit the conditions and situations found in Mexico City.

From Slaughterhouse to Corpse Deposit

Nonetheless, state officials struggled to mold the behavior of residents when it came to proper hygienic disposal of the dead at designated corpse deposits. By May 1905 the problem had become so pervasive that the state enlisted the help of Mexico City Police Inspector Domingo Martínez to find a suitable building that could store large numbers of bodies. However, the officials wanted the new building to be close to an existing rail line, a prime example of how important the railroad was to the Porfirian administration. Workers would load corpses from the building and corpse deposits onto railroad cars and take them away to protect public health.[94] On May 12, two days after receiving this request, Martínez believed he had found two buildings—both near rail lines—that could alleviate the city's corpse troubles. However, after further inspection, the police inspector realized that the buildings would require extensive repair in order to turn it into a storage unit for the dead.[95] So he continued to search for a building that was move-in ready.

Six days later, on May 18, Martínez found a building that met the needs of state officials. A large shed sat vacant on the grounds of a well-known city slaughterhouse named San Lucas, that had recently fallen out of favor with city officials; the main slaughtering operation had move to the renovated Peralvillo neighborhood. Thus, state officials viewed San Lucas as antiquated and out of touch with the demands of a growing population. Originally constructed in the eighteenth century, government officials had already targeted San Lucas several times for renovations, but by the turn of the twentieth century, they had abandoned any further attempt to renovate. Its colonial design, particularly the red volcanic rock used to make the floor, was porous and absorbed the blood and other liquids that gushed from slaughtered animals, which officials argued had turned it into a public health hazard.[96]

However, as a potential site for storing bodies, San Lucas seemed appropriate. Mexico City's late nineteenth-century population growth had sparked an increase in the number of working-class neighborhoods surrounding the slaughterhouse. The hygienic conditions in the nearby neighborhoods, lined with the homes of the urban poor, were as dismal as the slaughterhouse. In particular, the lack of hygienic facilities led the urban poor to engage in behaviors such as urinating or defecating in the street, which state officials and well-to-do residents considered backward and uncivilized.[97] If the poor wanted to act like animals, then what better location than a slaughterhouse was there for storing their carcasses?

Corpse Deposits, Corpse Carriages, and Public Health

Although the San Lucas slaughterhouse had presented state officials with an opportunity to fulfill their desires (and demonstrate to the urban poor that modernization was inescapable), they ultimately decided that the building "was not convenient to use" for storing corpses.[98] While officials gave no specific reason, a potential explanation emerged in July 1905, when they chose to construct two large corpse deposits inside Campo Florido and Los Angeles—both colonial-era cemeteries that the government owned. Despite having rejected a similar plan for corpse deposits inside Campo Florido in 1890, the corpse problem in the city had grown tremendously between 1890 and 1905, and thus the state required more burial space.[99] State officials soon settled on hiring highly respected military engineer Carlos Noriega to design and construct "hygienic, simple and economical" corpse deposits as well as provide corpse carriages for transporting the bodies from the overflowing deposits to the cemeteries.[100]

State officials had chosen Campo Florido and Los Angeles for the construction of Noriega's corpse deposits because of their remote locations and also because each cemetery had colonial ties that were useful for the Porfirian government's project to establish an official national identity. Additionally, Noriega's design for the deposits reflected a popular trend in architectural projects sponsored by the Porfirian state. For decades Mexican elites believed that Paris and the changes instituted by Haussmann in the mid-nineteenth century were the archetype of modernity, which they sought to emulate.[101] The 1889 World's Fair held in Paris, attended by over 28 million visitors, offered Mexico an opportunity to develop an official history for international consumption.[102] According to Porfirian officials, Mexico's exhibit at the world's fair had "to highlight the great, though atypical, lineage of the nation it represented: a national entity with a glorious past but ready to adjust to the dictates of cosmopolitan nationalism and eager to be linked to the international economy."[103] The result was the Aztec Palace, a building that cost the Mexican government about 280,000 pesos ($5.25 million) to construct and an additional 29,000 pesos ($544,000) to furnish.[104]

Moreover, the Aztec Palace enabled state officials to display in a safe and controlled manner their past, present, and future, all critical elements of Mexico's burgeoning national identity. Its architectural design was integral to the narrative: the palace needed to be a "building which at its sides and angles would characterize the architecture of the most civilized races of Mexico, but which would distance itself from the dimensions of ancient monuments that opposed modern amenities and tastes."[105] To be

Fig. 1. *Entrance to the Aztec Palace at the 1889 Paris Exposition.* Library of Congress, Prints and Photographs Division, LC-USZ62-102655.

a modern Mexican, at least according to both Porfirian state officials and elite citizens, one had to adopt a particular view of the past, highlighted by the fact that the country's rich Indian history was a distant memory. This was part of the continuing "long ideological and cultural Mexican tendency to selectively reevaluate the Indian past as part of the national identity," which reinforced the popular Porfirian notion that the only good Indian was a long-dead Indian.[106]

To demonstrate this vision for modern Mexico, state officials chose a design that had been submitted by historian and statistician Antonio Peñafiel. He and engineer Antonio de Anza based the building's exterior on a synthesis of pre-Hispanic architectural styles taken from a published book of Mesoamerican antiquities collections assembled by Irish antiquarian Edward King, Lord Kingsborough.[107] The building—a reproduction of a *teo-*

calli (an Aztec temple)—measured 70 meters (230 feet) long, 30 meters (98 feet) wide, and 14.5 meters (47.57 feet) tall. The construction of the temple, Peñafiel decided, had to reflect both Aztec architectural elements and modern, late nineteenth-century architectural standards, which would later serve as the foundation for his corpse deposit designs. For example, he decided to use a steel frame and a glass ceiling, along with Mexican-style columns (instead of traditional neoclassical Greco-Roman columns), which he had seen when he visited the pre-Hispanic archeological site of Tula, Hidalgo. The temple's façade depicted Aztec religion, agriculture, and arts in a linear fashion, from the beginning of Aztec civilization to its end, the starting point of Mexican nationhood.[108]

The material used to construct the Aztec Palace inspired Carlos Noriega's choice of materials for building the corpse deposits inside both Campo Florido and Los Angeles. He chose to combine traditional and modern materials. Indeed, the material used to construct the façades was representative of the country's indigenous past: its outside walls were a combination of *tepetate* (an indigenous volcanic rock) and sun-dried brick—both features of the country's traditional architecture.[109] The inside, by contrast, reflected the trappings of modern Mexican architecture: easy-to-clean cement for the inner walls, floor, and ceiling.

The use of cement in the construction of the deposits was just the beginning of Noriega's contribution to tangible modernity in Mexico. He also incorporated several hygienic features to protect both public health and the health of cemetery workers. A wooden door with two circular holes positioned in the center, one above the other, provided the only access to the deposit. Iron bars covered in wire fabric crossed each hole, which allowed fresh air to circulate inside the deposit. Additionally, Noriega believed the wire screen prevented flies from entering and laying eggs inside the bodies as well as rats from chewing on flesh, which had the potential to not only infect workers but also create a large-scale pandemic like yellow fever or rabies.[110] The design also featured seven semicircular windows that allowed natural light to enter. This combined with ventilation—both common characteristics of mid- to late nineteenth-century North American architecture—sought to protect public health and reduce the spread of disease, both important elements for modernizing Mexico City.[111]

Additionally, Noriega's design also demonstrated the common theme of the Porfirian era: combining the past with the present. Noriega designed new corpse carriages, powered by horse or mule, which like the corpse deposit maintained a traditional façade with a modern interior. The carriages used cylindrical rollers and individual storage compartments to aid efficiency and promote its hygienic possibilities. Cylindrical rollers facilitated the loading and unloading of corpses collected at deposits and delivered to cemeteries. The carriage interior also included a reinforced double wooden bottom to support the weight of the loaded carriage, usually between "900 and 1000 kilos for four corpses and their coffins." Underneath each compartment was a slide that pulled out to form a ramp, aiding the arduous task of loading and unloading coffins as workers had done in earlier years; workers exerted less energy this way, which meant they were able to work longer hours.[112] Moreover, four storage compartments for corpses offered protection for the health of works by maintaining the separation of individual corpses (one corpse, one coffin) unlike in earlier years when workers placed corpses into carts without coffins or attempted to stuff more than one corpse into the same coffin.

The new carriage design reinforced one of the principle ideas that comprised the Porfirian definition of modernization: scientific management. Integral to this type of management was worker efficiency, a concept popularized by Frederick W. Taylor's late 1890s time and motion studies.[113] As a principle, scientific management was concerned with the physical efficiency of workers, who Taylor believed "could be retooled like machines, their physical and mental gears recalibrated for better productivity."[114] Taylor's modern production techniques, historian E. P. Thompson has written, required the use of concepts such as time thrift

Fig. 2. Guillermo Olivares and Ignacio Huerta Campi, "Rear View of Funeral Car Model." The back of the cadaver car pictured (from 1918) contained a ramp (labeled 1) similar to the one Carlos Noriega had proposed in 1905. Archivo Histórico de la Ciudad de México, Fondo-Ayuntamiento/Gobierno del Distrito Federal, Serie-Panteones, caja 3472, expediente 260, August 30, 1918.

and work discipline; however, for these concepts to be effective, management would have to restructure the habits of workers.[115]

Thus, management practices began to emphasize standard, precise procedures for each laborer, eliminating decisions based on tradition or rules of thumb. Taylor's approach to making labor more efficient first appeared in 1898 at the Bethlehem Steel Plant in Bethlehem, Pennsylvania. He calculated that with precise movements, tools, and sequencing, each worker could load 47.5 tons of steel per day instead of the typical 12.5 tons.[116] Taylor called these changes "social efficiency," a concept that focused on using leadership characteristics employers had learned to mold the behavior of workers.[117] As a result, he argued, modern workers would begin to have the same interests as those of their employers. Noriega's apparent incorporation of these principles into his corpse carriage design was significant for two reasons.

First, it meant that he was an early adopter of the ideas that Taylor later codified. Second, the use of such a principle meant that specialists like Noriega were well aware of American technological accomplishments and were eager to put them to work in Porfirian Mexico.

Mexican state officials believed that substituting modern technology and scientific management principles for tradition were essential features of the modern world.[118] As a result, the Díaz government spent about 29,000 pesos ($373,000) for Noriega's sixteen corpse carriages, two deposits, and furnishings, an insignificant sum in the government's budget at the time.[119] For example, in 1898 La Compania Limitada del Ferrocarrilles del Distrito Federal (the Railroad Company of the Federal District), earned $80,000 ($2.39 million) from renting hearses out in the city and nearby suburbs for fees that ranged from 3 pesos ($41.20) to 140 pesos ($1,930).[120] The total investment between 1877 and 1910 in public works and communication infrastructure totaled about 1 billion pesos ($15.1 billion). The majority of the money came from foreign companies, which supplied 667 million pesos ($10.1 billion), while 286 million pesos ($4.3 billion) came from private funds and only 83.9 million pesos ($1.26 billion) from the government itself.[121] Foreign companies and capitalists invested heavily in Mexican economic infrastructure, controlling 67 to 73 percent of all capital invested in Mexico, which in the 1890s yielded investors a return of between 10 and 25 percent.[122] It appeared that investing in the modernization of Mexico was one of the best investments that money could buy.

Mexico presented foreign investors with an extremely lucrative opportunity, with full support from President Díaz and his state officials. Similarly, other costly public works projects undertaken at the time included the paving of Mexico City streets (8 million pesos [$121 million]), the construction of schools in the Federal District (2.5 million pesos [$37.7 million]), and the construction of the Monument to Independence in Mexico City (El Ángel de la Independencia) (1.5 million pesos [$22.6 million]).[123] While foreigners provided large sums of money for modernizing the capital, the Díaz government would spend some of its own money to invest in the construction of hygienic spaces and transportation methods for corpses to demonstrate to city residents that the modernization process in the capital was inescapable.

The Electric Tram

Between 1898 and 1910, another revolutionary and efficient technology helped to replace mule-powered trams and solidify the Díaz government's push to modernize the capital. Electric trams had already begun operating in the United States and western Europe, as well as in Latin America capital cities including Montevideo, Buenos Aires, and Rio de Janeiro.[124] In Mexico's Federal District some 10 million pesos ($118 million) of private capital (primarily foreign capital) financed the construction of electric trams. The company responsible for introducing this new technology was La Compañia de Tranvías de México (the Electric Tram Company of México), a foreign conglomerate led by American engineer Frederick Stark Pearson and financed by Canadian and European investors.[125] In April 1898 these investors purchased La Compañia Limitada del Ferrocarrilles del Distrito Federal, which had been the principal railway operator in the Federal District since 1897.[126] The purchase provided La Compañia de Tranvías de México with an opportunity to electrify an extensive network of railways and thus an opening to reap even more financial benefits from this new transportation mode.

Leading the charge to bring electric trams to the capital was American Frederick Stark Pearson. He had studied chemistry and mining engineering at Tufts University, and after graduating in 1879, he accepted a position as a chemistry instructor at the Massachusetts Institute of Technology. However, Pearson reentered Tufts and earned a bachelor's degree in civil engineering in 1883, a master's degree in electrical engineering in 1886, and a doctorate of sciences in 1901. After completing his master's, Pearson acted as a consulting engineer for many of the largest street-railway and power companies in the United States, Canada, Great Britain, Brazil, and Cuba. In 1903 the Mexican Light and Power Company—owner of all the electric light companies in Mexico City—asked Pearson to oversee the construction of an immense, forty-eight thousand horsepower hydroelectric plant in the city of Necaxa (one hundred miles away), which would carry electricity over high tension lines to Mexico City. These lines would supply power to four electric substations in Mexico City—La Nonoalco (in the colonial part of the city, near elite neighborhoods), Indianilla and Tlaxpana (in the northeast), and Churubusco (in the southeast); in later years, the same hydroelectric plant would supply power to two additional substations in Mixcoac (in the southwest) and Xochimilco (in the south).[127]

The station located at Nonoalco, or La Nana, as company officials called it, supplied power to a substation near La Alameda Park, where several corpse deposits existed. The electric tram not only allowed state officials to move residents through the expanding city faster than before but facilitated the transportation of the dead as well. As La Compañia de Tranvías electrified the railway tracks it had inherited (in some cases building entirely new tracks), state officials saw an opportunity to move corpses from deposits to cemeteries in a more efficient manner than previous infrastructure had allowed. The electric tram would become a symbol for Mexico City, part of "the makeup and perfume" associated with modernization.[128]

Public Health and the Electric Tram

Mexican state officials considered electric trams the future of transportation in the modern city. Nevertheless, there were major problems associated with them in the opinion of newspaper writers and the people who used them: the electric tram brought danger and death to the forefront of urban living. Articles constantly characterized tram drivers as dangerous, even referring to them as "*mataristas* (from the verb *matar*—to kill), a play on *motoristas*."[129] As a result, some upper-class residents in the capital did not want electric trams to continue passing through their neighborhoods collecting corpses.[130]

Much of the distaste for the electric tram had its roots in popular perception and reaction to the attention-grabbing headlines about railroad accidents in the 1880s and 1890s. While accidents impacted more railroad employees and innocent bystanders than onboard passengers, that did not stop reporters from painting a gruesome image of these wrecks. Most wrecks happened close to towns, villages, fields, and farms where people lived and worked.[131] This allowed reporters to construct a narrative to capture the public's attention regarding the horrific nature of this symbol of modernity. Rather than serving as a symbol of hope and a step toward progress as Porfirian state officials had envisioned, the railroad and subsequent similar technologies such as the electric tram became part of the "massification of death" in Mexico that emerged from the ease with which new technologies could kill.[132] As a result, technology as a technique for population management actually backfired; instead of creating a favorable impression among residents, it had created an impression that modernization put them on a collision course with death.

In March 1909 several families in the posh southwest neighborhood of Colonia Cuauhtémoc outlined their disgust with one of these electric trams. Nicknamed "La Gaveta" (the drawer), the tram was responsible for moving corpses through the city to their final destination. However, the electrified railway ran through an exclusive and fashionable neighborhood much to the chagrin of its residents. In their opinion, the tram constituted a threat to their health.[133] Colonia Cuauhtémoc included access to modern public services like piped drinking water, indoor plumbing, and electricity, along with sidewalks, large trees, green spaces, and wide streets named after famous European cities like Rome, Milan, London, and Berlin.[134] Its residents had paid to shield themselves from distasteful scenes found in the capital and to ensure their health. One of the neighborhood's most powerful residents, Hugo Scherer Sr., a well-known, influential private banker who had intimate ties to the Díaz administration, became the spokesman for the neighborhood.[135] The Ministry of Foreign Relations, for example, selected one of his houses to serve as accommodations for German dignitaries during the 1910 centennial celebration.[136] Scherer and his fellow residents believed that a tram devoted to death, as La Gaveta was, brought "a repugnant spectacle" to the streets of the upscale neighborhood and could lead to the spread of disease since the tram moved corpses "in various states of putrefaction" that had been "left for some time in the sun" near the official deposits.[137]

Despite the potential health risks associated with the passing tram, there was a simple explanation for why it had to pass in front of Colonia Cuauhtémoc. The route proved to be faster and more direct to Panteón Dolores—only five kilometers (3.1 miles) west of the neighborhood—than other options, which meant it more efficiently took corpses to burial in the city's largest public cemetery.[138] If the tram route had to go around the exclusive neighborhood, it would add valuable minutes and perhaps hours to the delivery time, which would prove to be far less efficient and thus less valuable to the Porfirian modernization project underway in the capital.

In an effort to shield themselves from perceived danger posed by corpses, residents of Colonia Cuauhtémoc contacted the Electric Tram Company to request that La Gaveta change its route to avoid the neighborhood. Despite the political influence held by these residents, exemplified by individuals like Hugo Scherer Sr., the Electric Tram Company refused to alter the tram's route to Panteón Dolores.[139] While the tram company did not offer an explanation as to why the path could not change, it is safe to presume that Porfirian state officials had a hand in the decision. State officials most likely wanted the tram to pass through Colonia Cuauhtémoc regardless of public outrage since it represented the most efficient and direct route for disposal. Moreover, Porfirian officials wielded tremendous influence, so if the tram company wanted to continue to operate (and benefit from the patronage of the Mexican state), it would do what was best for business: only listen to state officials, not angry citizens. While the perception that the tram posed a public health risk remained, the tram continued to operate and deliver corpses to Panteón Dolores as if nothing had happened. For the government, the electric tram

FIG. 3. Funeral Cars of Mexico, Part 1. *The Street Railway Journal*, March 1898, 131.

that went through Cuauhtémoc was hygienic and efficient and most importantly provided more tangible proof that the government remained committed to modernizing the city's landscape regardless of how offensive certain efforts appeared to residents.

The delivery of corpses to public cemeteries like Panteón Dolores was an important symbol for the Porfirian government. This helped to reinforce the population management technique of spatial separation according to class that state officials had worked hard to construct in the living world. Class was the primary factor in determining an individual's space in the cemetery, which was similar to how the government divided space among the living in the city. To apply this approach to the deceased would demonstrate ultimate control over the bodies of residents. Families of the deceased could choose among six class options for burial, each with different costs. Elite families

FIG. 4. Funeral Cars of Mexico, Part 2. *The Street Railway Journal*, March 1898, 132.

often selected first-class graves, where permanent burial was an option—as was the construction of grand monuments of remembrance to the deceased.[140] In each grave class (except the sixth), families could purchase the site for ten years or perpetuity. After ten years, cemetery workers would either remove the remains from the grave entirely (often providing medical students with the bones of the deceased in order to make their own skeletons) or dump the remains into the grave to make room for new burials. In 1887, for example, the burial fees associated with ten-year graves for adults in the first-class section cost 80 pesos ($1,680), and permanent burial cost 250 pesos ($5,080). As a result, such high prices for first-class graves at Dolores acted "as a subsidy for the free pauper burials."[141] Sixth-class graves, however, were free; in the sixth class, cemetery workers buried the corpses of the urban poor in mass unmarked graves near retaining walls in the out-

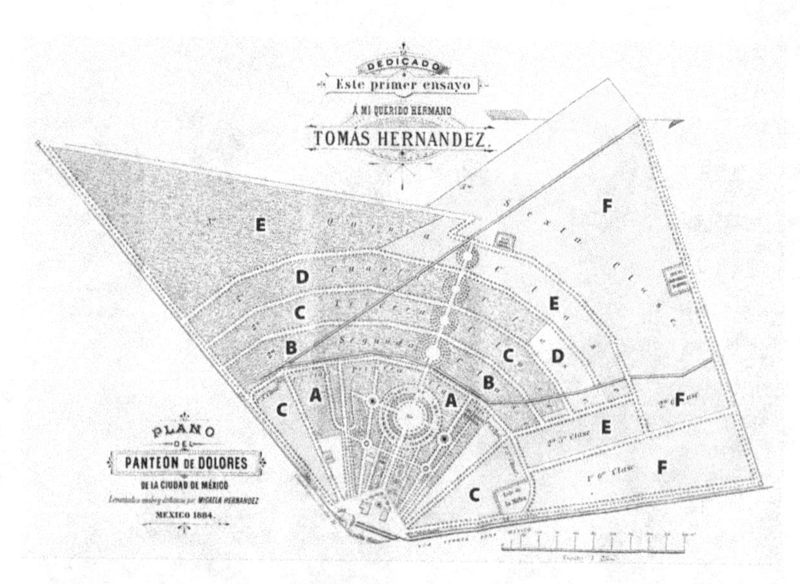

MAP 2. Micaela Hernández, Plano del Panteón de Dolores de la Ciudad de México, 1884. Courtesy of Nettie Lee Benson Latin American Collection, University of Texas Libraries, University of Texas at Austin.
Key: **A.** First-Class Graves, **B.** Second-Class Graves, **C.** Third-Class Graves, **D.** Fourth-Class Graves, **E.** Fifth-Class Graves, **F.** Sixth-Class Graves.

skirts of the cemetery, where grave maintenance was absent and long grass grew wildly, erasing the presence of the poor.

Despite their plans for safety being ignored by the government, residents of Cuauhtémoc remained steadfast in their commitment to change the route of the corpse tram. The first step that demonstrated their determination was to hire a lawyer, who they hoped would have more success at navigating the legal system: José Luis Requeña, a prominent Mexico City lawyer and businessman who shared an interest in the health of the neighborhood since he had built several homes there. Requeña had made his money from being the director and principal shareholder of Dos Estrellas Mine in the state of Michoacán. An American magazine described Requeña as a man who had ties to the largest financial institutions in Mexico but remained "one of the most level-headed men" in the country and "not a fawning sycophant" of President Díaz.[142] Thus, residents in Cuauhtémoc believed that he could use his political and social clout to bring the complaints about the tram to someone who could get the route changed.

Requeña contacted Governor Guillermo de Landa y Escandón, whom he knew well, as both were part of the same small social circle shared by prominent politicians and men of Mexican society.[143] On May 13, 1909, Requeña wrote to the governor, telling him that the tram passing through Cuauhtémoc was "dangerous and contrary to the principles of elemental hygiene" and demanding that it be rerouted.[144] His social connections worked. Less than a week later, the governor proposed a plan that would keep "the cadavers of the poor" from threatening the health of the Cuauhtémoc neighborhood.[145]

The governor's plan called for the Department of Public Works—responsible for overseeing the corpse deposits located throughout the city, among other duties—to find land near Colonia Cuauhtémoc where workers would construct yet another corpse deposit. This deposit, Guillermo de Landa y Escandón ordered, had to be in an area where electrified railroad tracks already existed in order to for the tram to avoid passing through Colonia Cuauhtémoc. For the governor, the health problems associated with the tram were now paramount: this change "was the best way to avoid the serious evil" associated with the presence of the putrefying bodies passing by some of the city's most elegant houses and residents. The governor offered a viable solution to the problem. Changing the tram's route to avoid the privileged areas of the city would keep the elite safe from any diseases emitted by the "cadavers of the urban proletariat."[146] The plan would also attempt to conceal the realities of urban life in the capital. However, in doing so, the government's attempt to delineate neighborhoods according to class backfired as the city's growing population combined with limited space for both people and technology meant blurred boundaries. The government wanted to present residents and visitors alike with an officially constructed version of Mexican modernity, highlighted by the continued use of the latest technology and hygienic protocol.

Combining Tradition and Modernity

The Department of Public Works could not find space in the city to bring the governor's plan to fruition. In July, Manuel Escalante, the department's director general, admitted that he and his workers had been unaware of any significant problems associated with the corpse trams, as "it has been some time since we studied the free service of collecting the cadavers of the poor."[147] However, in an effort to appease the governor and demonstrate solidarity with Porfirian officials, Escalante assured Porfirian officials that his team would develop its own plan to keep corpse trams from threatening the health of the city's exclusive neighborhoods.

Escalante's proposal combined both traditional and modern methods of corpse collection. Instead of allowing electric trams to pass through elite neighborhoods like Colonia Cuauhtémoc, the department would use a traditional collection method, four mule-drawn carts, to collect the bodies of the urban poor from nearby deposits. Workers would then load the corpses onto the carts. Once the carts were loaded, workers drove the carts to predetermined locations in the city where electric tram tracks existed in order to deliver the corpses to Panteón Dolores. Escalante's plan also called for state officials to find an existing building or piece of land that the Department of Public Works could modify for emergency corpse storage—such as during an epidemic, when the number of dead urban poor would surely explode.[148] However, Governor Guillermo de Landa y Escandón rejected this plan. He explained that there "were no funds available to cover the expenses," and even if there were, it was "not appropriate to bring new spending" despite the plan's relative low-cost and therapeutic properties.[149] But the governor would not have to wait too long before he received a better offer for transporting the city's corpses.

No $

Completely Modern

In late July 1910 Luis Riba, the general manager of the Electric Tram Company of Mexico, presented the governor with a new and modern solution to the city's corpse transportation problem. In a letter to the governor, Riba shared some exciting news: the Electric Tram Company was now capable of making corpse transportation exclusively electric. He argued that mule-drawn carts had become ineffective since they had to cover extensive areas of the city when collecting bodies, which meant the company had to spend more money to repair the carts and purchase new mules.[150] If electric trams replaced all mule-carts, not only would it make moving through the city faster, but the increased appearance of electric lines would surely help to present the capital as more modern. Moreover, the electric tram would become the sole method of corpse transportation in the city, and as Riba pointed out, it would protect public health by providing "conditions that can be fully disinfected, which will not allow for any infiltration of stench from the corpses."[151] Adopting the electric tram as the only viable method of corpse collection and delivery would be advantageous since additional miles of track were added to city streets each day.[152]

On August 3, a few days after the governor had received this proposal from Luis Riba and the Electric Tram Company, a letter landed on his desk from Aurelio Macias, the administrator of Panteón Dolores. Macias had heard that the Electric Tram Company had presented the governor with an opportunity to use electric trams as the exclusive method for the collection and delivery of corpses. Yet, Macias had heard—probably from someone at the Electric Tram Company, since both the company and the cemetery would benefit—that the governor was only willing to pay for the use of one tram for the entire city. This was a problem. Macias pointed out to Governor de Landa y Escandón that even two mule-drawn carts could only carry eight corpses each, which meant a total of sixteen corpses being moved at once. Similarly, one electric tram could hold ten corpses, but as Macias explained, this was "insufficient for service given the number of cadavers found daily." But if the governor committed to using two electric trams, it would bring the total capacity to twenty corpses, making the collection and delivery of corpses more efficient and "much faster" than mule-drawn carts.[153]

The governor agreed. So on August 12 Governor Guillermo de Landa y Escandón told Luis Riba that the city wanted to convert its corpse transportation to electric trams only. After speaking with several people, the governor had decided that one tram was "too insufficient due to the number of corpses collected daily."[154] If Riba could provide two trams, then de Landa y Escandón would make the Electric Tram Company the city's official corpse carrier.[155] The relationship would benefit the government and the Electric Tram Company. For the governor, the use of electric trams in the collection of corpses would showcase his support for modern technology and his commitment to improving public health. For the La Compañia de Tranvías de México, it would provide them with a direct line to financial support and the potential for future projects through the office of the governor or maybe even President Díaz. Thus, the Electric Tram Company accepted the governor's counteroffer and sent two trams to begin collecting corpses in the city.

Leading state officials and elite residents in Mexico City were not the only people who supported the construction of new tracks and the use of the electric tram as the official method for corpse collection and delivery. On August 4, 1911, well-to-do residents living near the military hospital and municipal cemetery in Churubusco—"a straggling suburban town" in the Federal District, southwest of downtown Mexico City—also argued that if electric trams came to their town, their lives and public health would improve tenfold.[156] Writing to new governor Ignacio Rivero, the residents expressed anger over the military hospital's current method for corpse transportation. Unlike the modern method used in the capital, Churubusco's approach was stuck in the past and unbecoming of a modern nation. Hospital attendants stacked corpses in wheelbarrows in order to deliver them to the local municipal cemetery for burial. Sometimes bodies arrived in "poorly made boxes," leaving "a trail of blood" on the ground behind them. Residents complained that this method had created "a truly pathetic and disgusting spectacle" that "endangered public health." However, the residents believed they had found a solution that would "remedy the evil" from their neighborhood and showcase the suburban's town commitment to modernity.[157]

Churubusco's citizens asked Governor Rivero to get the Electric Tram Company to send a tram to the military hospital to collect the dead. Electric tracks already existed in the city that linked it with the capital, and thus the same tram used in Mexico City to collect and deliver corpses could come to Churubusco "with the same object." All the governor had to do to make this a reality was have hospital employees in Churubusco notify the tram operator by telephone—another symbol of modernity—that there were corpses ready for pickup. If there were no bodies to collect, hospital employees would call the Electric Tram Company to tell them not to come. However, if there were corpses that required pickup, the same employees would phone the tram company to report how many required transportation. Residents believed that using both the telephone and the tram would "provide great savings" financially for the government, generate goodwill between the state and city residents, and allow modernity to extend into the appreciative suburbs.[158]

It appears the government listened to the residents of Churubusco. In December 1912 the governor instructed the Department of Public Works to begin to send trams to the military hospital in order to return with corpses for burial in Panteón Dolores.[159] The electric tram had begun to cement itself as an important symbol in the elite and middle-class discourse on modernity. Electric trams that collected and delivered corpses in Mexico City and nearby suburban towns allowed the government to provide additional proof to its citizens that modernity was inescapable in the Federal District. Furthermore, by focusing on the corpse dilemma, state officials could avoid having to provide solutions for improving individual health and reducing disease; they could eliminate the spread of many diseases by introducing sanitary surveillance methods for the entire population rather than focusing on the lower classes as those susceptible and responsible for the potential spread of disease.

Conclusion

Despite the supposedly positive changes made to the transportation of corpses during the Porfirian era, even in January 1918 aspiring funerary automobile entrepreneur Rafael M. Peña remained one of many citizens who continued to lament the poor state of corpse transportation in Mexico City. Peña wrote to Federal District Governor Alfredo Breceda to tell him that he and countless other residents still considered the electric trams used for the transportation of corpses over the past eight years to be a danger to society. In particular, he told the governor that corpses often traveled on trams in crudely constructed coffins, many of which went uncovered entirely. Thus, what state and city officials thought was a method of protecting public health had become a threat; public health and hygiene were at risk in the city. The solution, Peña argued, was for the city to enter into a two-year contract with his funerary automobile business. He promised the governor that not only was his system "modern, elegant, and practical" but that once it was adopted, the governor would see immediate improvements in the hygienic transportation of bodies and public health. His proposal would also help the governor save money; Peña guaranteed that he would charge the city a much lower rate for transporting corpses than the La Compañia de Tranvías de México did.

But being modern and affordable was just the beginning, according to Peña. In particular, his transportation option was more valuable because it "would not pose any danger of contagion in the streets." His would only use the less traveled streets in the city, away from elite neighborhoods and populous areas, to arrive at Panteón Dolores. Peña also explained that the old method was inefficient since the routes often used popular streets, which meant obeying traffic laws and pedestrians and therefore delayed arrivals. The result, he explained, was an increase in the potential that disease could escape the coffins and endanger public health.[160] He concluded his letter by telling the governor that this problem had a solution: if the governor would pay him between 200,000 and 300,000 pesos ($6 to $9 million) for his automobile service, it would protect public health and help to shape the modern image of the city that state officials desired.[161] Adopting the automobile as the primary transportation method of corpses would align with the beliefs of state officials and elite residents who wanted to commit to using the latest technologies in order to showcase that the city was continuing to take the right steps toward modernity.

Two months later, on March 24, 1918, the governor directed a Department of Public Works employee to respond to Peña's request. The response was positive. The department saw no problem with establishing a corpse transportation business as long as it followed the hygienic transportation rules established by the Superior Health Council. Yet there were a few details that the governor wanted addressed. Interestingly, one of these was the cost of his service—despite the price having been listed in Peña's proposal—which perhaps was the governor's polite suggestion that Peña's prices had been too high. The proposal itself, however, remains intriguing for two reasons. First, it facilitated the introduction of four additional requests for establishing funerary automobile services that occurred between January 1918 and December 1919.[162] Second and more important, Porfirian officials' interests in controlling the transportation of the dead had cemented itself into the fabric of the modern Mexican state, leaving an important and (un)intentional legacy surrounding death and its role in the modern world.

In the Porfirian era, an essential component of this process was the opportunity presented to state officials to mold how city residents interacted with their environment. For Mexican state officials, how the living used the space around them contributed (positively, negatively, or both) to the image of the capital. By promoting the latest modern innovations, such as the railroad, sewers, indoor plumbing, electricity, or even safer disposal methods for corpses, state officials believed that progress could appear at a faster rate than in previous years. All residents in Mexico City would soon walk, talk, eat, greet, or mourn each other in a uniform manner. Integral to achieving this goal was the use of technology, which would serve as a way to eliminate the myriad differences among the residents of the capital, who came from vast social, economic, and cultural backgrounds.

Technology for transporting corpses presented state officials with a unique opportunity to exert their power over death in both intangible (emotions) and tangible (burial process) ways. By controlling the movement and disposal of dead bodies within Mexico City and, officials hoped, other towns in the Federal District, the Mexican government and its ardent supporters believed that each of the latest technologies they used would bring them one step closer to becoming modern. The 1910 version of the capital would represent a significant contrast to the capital of 1905, and the capital in 1905 was quite different from the 1900 version, and so on. The use of hygienic transportation for the dead had become part of the city's urban symbolism. Rather than a simple reflection of society, urban symbolism—here, the transportation methods for the dead—exposed how state officials used transportation in a strategic manner to shape and change social relationships in the city. These officials tried to create important mechanisms that would pressure all citizens, especially the urban poor, toward uniformity and social cohesion.[163]

In the end, anyone or anything left in the capital would be subject to the obsession over the question of how best to achieve modernity. Starting with Porfirian state officials and continuing with revolutionary state officials, the entire Mexican nation would benefit from the capital becoming modern. Accordingly, the processes that had unfolded in the capital over two decades were essential components of the paradigm of modernity, which all cities and towns in Mexico needed. This approach to state building allowed Mexican state officials to distance themselves from the past—a tumultuous nineteenth century before Porfirio Díaz that historians have characterized as chaotic and unstable, though normal for a newly independent countries—and never look back. The only thing that mattered was what lay ahead on the path to progress.

2

"An Extraordinary Tool"

Building a Modern Public Health System through Anatomical Dissection

During the second half of the nineteenth century, there was a push throughout North America to create a scientific alliance between Canada and the United States. Health officials, medical doctors, and scientific organizations and institutions used this alliance to focus on improving public health and sanitation in both countries in an attempt to foster a spirit of pan–North Americanism. Nevertheless, their neighbor to the south, Mexico, remained absent from the alliance. So the American Public Health Association's (APHA) invitation to Mexican medical professionals to join its influential international organization as a new member country in 1889 represented a tremendous shift from Mexico's earlier exclusion to now conversations taking place about how Mexico, Canada, and the United States could work together to address their shared public health and sanitation concerns.

Mexico's late addition to the alliance was the result of two important events in the country's modern history. First, President Porfirio Díaz's second term began in 1884 and brought with it stronger recognition and economic investment from abroad and at home. Díaz and his state officials began to receive unwavering support from middle- and upper-class citizens to improve public health throughout the country in ways that would continue to help Mexico maintain its newfound alliance. This strategy was one that many medical professionals and state officials believed could be instrumental for preserving the Porfirian push to modernize. Additionally, this forged alliance masked the true intention behind Mexico's inclusion. Mexico's incorporation into the APHA had more to do with protecting American, Canadian, and other countries' workforces from diseases such as yellow fever, tuberculosis, and diphtheria, which sanitarians believed to be specific to "backwards" and "barbarous" nations such as Mexico.[1]

Nevertheless, Mexican physicians and sanitarians did not let these characterizations stop them from embracing the potential success this new relationship and emphasis on public hygiene could bring the country. At the 1890 annual APHA conference in Charleston, South Carolina, the address on sanitation delivered by association president Dr. Henry B. Baker of Lansing, Michigan, explained that in civilized society "life and health of every person is more or less bound up with the life and health of every other person, that not only is man his 'brother's keeper,' but on each person there rests some responsibility for the welfare of all."[2] This idea resonated with Mexican medical officials, who like President Díaz saw themselves as the saviors of the people of Mexico. Improving public health through a combination of population management and hygienic science presented them with an opportunity to improve Mexico's reputation and the public good.

One individual who attended this annual meeting and who would be instrumental in recognizing its potential for modernizing the Mexican "medical profession and its institutional structures" was Dr. Eduardo Licéaga.[3] Chosen in 1885 as the president of the Superior Sanitation Council—a council of the Federal District's Board of Health that focused the majority of its efforts on Mexico City—Licéaga was the man responsible for overseeing much of the city's efforts to improve public health. In particular, the sanitary conditions found in state institutions such as hospitals, jails, and cemeteries required his immediate attention. As an organization, the SSC had almost unlimited power for solving public health problems, including intervening "when it thought it was necessary to do so" as well as asking other governmental departments for "any information it needed for its work."[4] However, beginning in the early 1890s, Licéaga used his position and influence in the SSC to address what he considered the root of the city's public health problems: inferior medical education.

For Licéaga, Mexico City's public health conditions had direct ties to the training that physicians had received in recent years, which he considered to be substandard. Thus, the National School of Medicine located in Mexico City—the nation's most influential medical school and Licéaga's alma mater—became the focus of his efforts to improve the situation. Licéaga felt that after he had graduated two decades earlier, university administrators and professors had abandoned the practical and invaluable experience of dissection in the curriculum. So from the very beginning of his term on the council, he focused exclusively on improving teaching and learning. He stressed the importance and value dissections had for improving students' understanding of the human body, which he believed would yield improved results in public health and simultaneously demonstrate Mexico's commitment to modernization.

Licéaga was able to devote a significant portion of his professional career to improving the reputation of Mexican medicine because he was President Porfirio Díaz's personal physician. Their close relationship allowed the two to work intimately "to build a formidable public health apparatus in Mexico" and a modern medical curriculum.[5] Additionally, Licéaga benefited from Díaz's commitment to modernizing the capital (and the country) through science and technology—embodied by the passionate relationship that existed between medical professionals and state officials—to solve the public health problems that plagued the Porfirian administration.

In this chapter, I examine how changes to the medical curriculum at Mexico City's preeminent medical school, the National School of Medicine, intertwined with the government's desire for population management through improvements in public health. With the help of President Díaz's personal physician,

Eduardo Licéaga, the university worked closely with state officials to improve the health of citizens by promoting a medical curriculum that focused on anatomical dissection and hands-on training as a way to solve much of the city's public health dilemmas. Licéaga and other medical and state officials also believed that the existing curriculum relied too heavily on rote learning, which had inadequately prepared physicians to think on their feet when dealing with real world challenges, in particular, the overwhelming amount of sickness and disease found in Mexico City.

I argue that the changes Licéaga would make to medical education occurred over three periods: 1889–97, 1898–1905, and 1906–11. Licéaga himself never declared distinct periods for these changes. However, in examining the university records, it becomes clear that each of these periods was best characterized by the growing importance placed on dissection as a means for improving the physician's understanding of the problems found inside the body, which was an integral part of improving public health. Furthermore, these changes overlapped with the Porfirian desire to manage the population by encouraging both everyday citizens and medical professionals to embrace the results an improved understanding of the human body could provide and accept medicine for what it was becoming: the religion of modern man.

An Extraordinary Tool

Every morning, physician Don Ramón Fernández looked outside his apartment window onto the busy streets in front of Plaza Santo Domingo, a bustling neighborhood for middle- and lower-class residents of Mexico City.[6] From the same window, he could also see the National School of Medicine, located in the former headquarters of the Spanish Inquisition. On the school's roof, medical attendants known as *mozos* busily sorted partially dissected body parts into large piles designated by type (arms, legs, torsos, etc.). Storing these parts on the roof offered faculty, staff, and students a convenient solution to a problem that had existed for years and one that would plague the school until the introduction of special cadaver refrigerators in 1912.[7] Instead of storing body parts inside a room at the medical school where miasmatic theory—that disease was caused by the presence of miasma, a poisonous vapor caused by particles of decaying matter, remained a popular explanation for how disease spread—the roof presented an alternative storage space that reduced exposure to potential miasmas. Exposure to the weather of Mexico City meant accelerated rates of decomposition and more importantly kept the gruesome aftermath of dissection away from students, professors, and potential visitors.

Yet Don Ramón Fernández was not pleased by the view he saw from his window. The landscape of cadaveric remains offended his sensibilities and upset his daily routine to such a degree that he believed in order to fix the problem he had to do something himself. So he wrote to Dr. Manuel Carmona y Valle, the director of the medical school, to complain about the stressful sights he endured daily and offered to pay for the construction of a wall on the roof of the school to obscure his view of the decomposing remains.[8] However, the school dismissed Fernández's offer. Carmona y Valle instead sought the help of the minister of justice and public instruction—who oversaw all public schools, including the National School of Medicine—to explain to Ramón Fernández that the storage of dissected remains on the roof would continue.[9] The National School of Medicine, the minister explained, was a public institution that required "the most perfect independence" to achieve success.[10] The school understood Fernández's dismay; however, improvements in dissection at the school that could lead to improvements in public health outweighed his disgust. His opinion was one of someone out of touch with the modern world. Carmona y Valle further explained that if the school yielded to the demands of local residents, "it would seriously harm the progress of the youth educated there."[11] This explanation and pedagogical approach represented the beginning of the new direction that medical professionals and state officials would take when it came to medical education.

Furthermore, the case of Don Ramón presents us with an interesting predicament for state officials. How did Porfirian state officials feel about the impact the presence of corpses had on the state's construction of modernity in the capital? It seems as if the daily occurrences of complaints about dissected body parts or dead bodies found in the street were not considered impediments to achieving the desired progress. The case of Santa Paula (discussed in chapter 1), which took five years to complete, illustrates this point. Bodies arrived at the cemetery gates on a daily basis, removed from local deposits, in plain sight of residents and tourists. In Don Ramón's case, it was as if university and state officials believed that as a physician, presumably from an older generation, he should have welcomed the changes occurring at the school. The shift from rote to tactile learning was the key to progress. Improving the state of dissection in the medical curriculum was the only way to train first-rate physicians. In particular, dissection was an extraordinary tool: one that could simultaneously enhance the understanding of the body and help provide solutions to the city's myriad public health problems.

For many Mexican state officials and medical professionals, improving public health hinged on being able to make substantial changes to the curricula at medical schools. But before implementing wholesale changes at all medical schools in the country, Licéaga and other medical professionals worked closely with the administration of the National School of Medicine, which served as an incubator of progress. Thus, if the changes worked there, similar changes were implemented across the country. According to Licéaga, despite the increased number of professors and medical assistants found throughout the country since he had received his medical degree in January 1866 under the rule of Emperor Maximilian, Mexican medicine "had not evolved in 25 years."[12] This failure to evolve was due to the country's chaotic nineteenth century, marked by constant governmental instability that failed to place enough importance on improving medical knowledge and safeguarding public health; as a result, the curricula at Mexican medical schools fell far behind those in the United States and Europe.

Mexico's instability, which included a revolving door of presidents during the nineteenth century, had failed to invest in improving medical education and public health. Porfirian officials believed that past officials had overlooked the importance of dissection, which contributed to Mexico's inability to serve as a paradigm of modern medicine in North America. But when Porfirio Díaz became president, attitudes changed. He and his team of científicos (medical professionals, bureaucrats, lawyers, and engineers) focused their attention on creating tangible examples of modernity like establishing hygienic guidelines for corpse transportation on board trains in order to improve public health and the image of the capital. Thus, university and state officials saw an opportunity to create a new era for Mexican medicine by emphasizing the important role dissection should play in the training of the modern physician.

But for medical knowledge to improve, students needed cadavers to practice their skills. The use of cadavers in European and U.S. medical schools had long histories. From the beginning, the bodies of criminals and the bodies of the poor supplied medical schools, and when supplies were low, other avenues for acquiring corpses, such as graverobbing, emerged. Yet with the enactment of 1832 Anatomy Act in England, official channels of corpse distribution appeared as the state gave medical schools and anatomists unrestricted access to the corpses of the poor. In the United States, however, dissections did not begin until after the Civil War. Nevertheless, British and American students with money chose to travel to continental European cities like Paris, Leipzig, or Vienna, where there was little to no outcry over the use of corpses.[13]

While the act of dissection was not a foreign concept to Mexican physicians and medical schools, it was, however, according to Licéaga, an insignificant component of the curriculum at the National School of Medicine. Textbooks and medical theories held significantly more value. In Licéaga's opinion, these antiquated methods of instruction had led to students demonstrating incompetence when it came to "anatomy demonstrations and dissection exercises."[14] He also believed that the quality of students enrolled in the university had contributed to the unfortunate state of medical education. The majority of students, he wrote, were of "common aptitude" and those with "exceptional mental faculties were rare, exclusively rare."[15] The best and the brightest, many of whom came from well-to-do families, could afford to leave Mexico for Europe, a learning environment considered by most medical professionals to be of a higher caliber. In order to keep the brain drain to a minimum, the National School of Medicine had only one option for improving the state of medical education: reintroduce dissection.

Licéaga wanted the medical curriculum to reflect the importance dissection should hold in modern Mexico. In his opinion, theoretical studies had no value when compared to the experience that students could gain from working with anatomical material; as he explained, "performing dissections or executing operations on a cadaver cannot be replaced with dissections in a book."[16] According to Licéaga, the biggest problem in Mexican medical education was the lack of practical training offered to students. Instead, many of the classes involved students memorizing medical theories and organ layout from textbooks, which Licéaga considered antiquated instructional methods in the modern world. Theory was not a worthy substitute for the hands-on approach Licéaga valued.

To demonstrate his commitment to ushering in a new era in medical education, Licéaga presented the National School of Medicine three specific suggestions for improving its reputation and the state of its medical education. First, the National School of Medicine needed to collaborate with local hospitals like Hospital Juárez or San Andrés, since they had "airy wards with lots of light and sufficient ventilation" that would present students with the appropriate environments to practice their dissection skills. Second, the National School's classrooms and laboratories lacked the proper equipment necessary for students. As Licéaga wrote, "the furnishings of the departments are very far from having the conditions that exhibit modern pedagogy."[17] If the government was serious about fixing medical education, then the university needed to not only provide ample opportunities to dissect but also the latest medical equipment, such as anatomical mannequins, laryngoscopes (endoscope for examining the larynx), and spirometers (apparatus for measuring the volume of air produced in the lungs), all of which were standard in American and European medical schools.[18]

Furthermore, Licéaga used his last suggestion to drive his point home regarding the antiquated nature of Mexican medicine by criticizing the government's investment in medical education. The government was investing millions of dollars to construct public works and infrastructure projects, such as sewage systems and prisons, but the university—and medical education in general—received far less.[19] In order to improve public health, much of which could happen through updating the curriculum at the National School of Medicine, Licéaga wrote that the government had to spend more money "than what the Nation currently dedicates to the teaching of medicine."[20] If the goal for state officials was to turn Mexico City into a modern capital—with a first-rate medical education system—then they would have to pay for the associated costs that came with such a goal.

He also believed that the Mexican government paying for student tuition at the National School of Medicine was a problem. For Licéaga, this approach was a "burden that has been voluntary imposed" to foster a culture that valued education. However, it also limited the funding the medical school could receive, and consequently "the needs of medical teaching" remained unmet.[21] He suggested the government adopt the system in place at two of the world's preeminent medical schools—the University of Paris and Harvard University—where students paid for their own medical education, an approach dated back to the late sixteenth and early seventeenth centuries. According to Licéaga, Harvard charged students $5 for tuition, $200 for teaching fees, $30 for exam rights, $6 for entrance into dissection rooms, and $4 for instruments—$245 per year per student ($7,130 in 2017).[22] Adopting a similar approach for medical schools in Mexico, he thought, would immediately free them from the problem of insufficient governmental funds and lead students to value their education more since they, not the government, would be responsible for paying for it. Furthermore, Licéaga believed this approach "would establish a noble emulation between physicians who dedicate themselves to teaching" and students, who would be put in the "brilliant position to learn from and consult with medical authorities."[23] Such change in the medical education system would contribute to the professional and intellectual development of both teachers and students, which in turn would help improve public health in the capital and support the modernizing vision that state officials held.

While the government did not adopt Licéaga's tuition suggestion, state officials did listen to his ideas about how to *improve* medical education. In 1894 Joaquín Baranda, the secretary of justice and public instruction, worked with administrators and professors, including Licéaga, who was a professor of surgical therapeutics at the time, to modify the curriculum. In particular, the changes would reflect the important role anatomy cadavers would play in training modern physicians. Training at the National School of Medicine took a total of six years to complete, and for the first time in years, courses would no longer exist in isolation but serve as complementary forces, each emphasizing the importance of the tactile experience dissection presented. For example, students would take classes like descriptive anatomy in their first year, topographical anatomy in the second, pathological anatomy in the third, and surgical pathology in the fourth. By the fifth year, they would attend and perform autopsies at the school and local hospitals. Ultimately, in their sixth and final year of medical school, students would be so familiar with anatomy that they would spend the year working in medical clinics to study how diseases and other illnesses affected living patients. With so much of the new curriculum focused on demonstrating tactile skills, Baranda had to change the way professors tested students. Instead of making students produce written essays as their final product in classes, they would now have to perform a dissection. Furthermore, one-third of the mandatory professional exam required to graduate included a dissection component as well.[24] The importance placed on dissection represented a fundamental shift in the direction of medical education at the National School of Medicine. State officials and medical professionals believed that by emphasizing dissection, the university's archaic medical pedagogy and curriculum would improve tremendously and deliver results that would be invaluable for improving public health in the capital.

The Professional Anatomy Museum

However, dissection alone would not be enough to turn the National School of Medicine into Harvard or the University of Paris. What separated the best medical schools in the world from average ones was not only the emphasis on dissection in the curriculum but also the presence of an anatomy museum. Here, students could enhance their studies by familiarizing themselves with historical and contemporary anatomical collections. But Mexico's most prestigious medical school did not have such a museum. However, in January 1895 a National School of Medicine professor named Sánchez submitted a proposal to President Porfirio Díaz that sought to change this situation. According to the proposal, building a museum at the school in Mexico's capital was "a necessity of vital importance;" an anatomy museum would be an "immemorial institution" that would put Mexican medical studies on the same level as other countries of "the first order."[25]

Anatomy museums were essential components of nineteenth century medical education. Universities used them as a way to attract students and acquire financial support from the government or wealthy patrons interested in improving the health of citizens through improved training of aspiring physicians. The professional anatomy museum itself was a repository for both typical and odd medical souvenirs. It contained "stuff in jars, skeletons, dried preparations, casts, and models in wax, plaster, papier mâché, and wood." But unlike the other anatomy museums also popular during this era, which were noted for their desire to attract the public, professional museums only allowed entry to physicians and medical students and on certain occasions to presidents, foreign dignitaries, or high-ranking state officials as well.[26]

These museums had become so integral to medical school culture that the condition of schools without them, wrote one medical instructor, were like "the state of man without language," a frightening characterization for Mexico.[27] As Sánchez pointed out, "there is no important medical center in Europe or America that does not possess a museum of this nature."[28] Museums provided an indispensable type of education by allowing medical school faculty to use artifacts to help bridge the past with the present. Like the new curriculum that focused on dissection, the creation of a dedicated museum had the potential to contribute to the "production and progress of knowledge" among the student body.[29] In addition to serving as a pedagogical tool, anatomy museums and collections were also important for state officials around the world, including in Mexico, because they helped to reinforce the message that science was an unmatched weapon in the fight for improved living standards and public health.

Moreover, the museum and its collections legitimized the medical profession and the rising social position of physicians. For Mexico and for the author of the proposal, Sánchez, the museum offered proof that medicine and the individuals who studied it had access to a specialized knowledge. In particular, they understood the "character of sicknesses that present themselves and the rest of the peculiar circumstances" found inside the human body. Sánchez believed that medical studies would help students receive accolades and respect from everyday citizens, foster a patriotic fervor among citizens, and "demonstrate that here [in Mexico] we cultivate science."[30] The anatomy museum would demonstrate Mexico's commitment to its burgeoning scientific reputation, to modernity, and above all to its citizens.

The museum would also help improve the reputation of the Mexican physician. According to Sánchez, people routinely threw out "unjust accusations against the most accredited physicians of our country" despite the accolades medical professionals received from distinguished international medical organizations like the APHA. Nevertheless, he did admit, as Licéaga had, that recent Mexican physicians had suffered from poor medical instruction and that anatomy departments had failed "to provide premium indispensable material." As a result, Sánchez proposed establishing a cadaver department—affiliated with the museum but located inside Hospital San Andrés near Alameda Park—that would provide physicians and students with the opportunity to practice dissection and create their own anatomical collections for the museum. Putting the cadaver department inside the hospital would provide students and physicians access to a significant number of unclaimed bodies, frequently of the urban poor, which would be instrumental for achieving Sánchez's vision of having three thousand to four thousand anatomical specimens cataloged.[31] Using the bodies of the urban poor who died in state hospitals was nothing new in Mexico or Europe. These bodies were important for Mexican physicians because unlike those found at corpse deposits throughout the city, these came with medical histories, which meant they offered an additional layer of protection from potential diseases than those found on the street.

Creating an anatomical collection of such magnitude would require significant financial support from the government. Sánchez projected that the museum would need a hefty monthly budget of 5,292.80 pesos ($699,000) to assure success. The museum would allot roughly 8 percent, or 440 pesos ($58,900), of the budget per month to cover employee salaries. The distribution for high-ranking employees was as follows: the chief physician responsible for overseeing the collection and preparation of anatomical pieces would receive 100 pesos per month ($13,200); a histology physician in charge of microscopic preparations would receive 80 pesos ($10,600); and a physician in charge of collecting and compiling statistical information about the clinical histories of cadavers from Hospital San Andrés would receive 50 pesos ($6,600). But the museum also needed to hire low-ranking employees to perform undesirable work. For example, a dissection room assistant, who would clean up after physicians and students performed dissections, would earn 15 pesos ($1,100) per month. The budget also called for equipment and materials, such as chemicals for preserving anatomical specimens, which would cost 80 pesos ($1,230), laboratory desks for 10 pesos ($154) each, and the containers that held chemical preservatives for 25 pesos ($384) each.[32]

While these costs may seem high, having such a museum was important for improving the status of Mexican medical education. The presence of such a museum offered state officials hope that talented Mexican students would stay in Mexico rather than leaving for well-known medical schools in the United States or Europe that had renowned anatomy museums. Students who could afford to go abroad or who received scholarships often chose to leave Mexico. Additionally, the country had a difficult time attracting students from abroad. It did not help that influential foreign physicians like Herbert J. Hardwicke—a member of the Royal College of Physicians and fellow of the Royal College of Surgery at Edinburgh, Scotland—had published scathing critiques of medical universities in Latin America. According to Hardwicke, Chile, Brazil, Argentina, and Venezuela all had "good universities and tolerably good medical laws."[33] However, outside of these four countries, he wrote, "the condition of medicine is as bad as can well be imagined."[34] In particular, Mexico received harsh criticism as a country that lacked uniformity in its medical regulations and its procedures for obtaining medical licenses. All of the five medical schools in the country—the University of Oaxaca, University of Campeche, University of Zacatecas, University of Guadalajara, and the National School of Medicine—were "in a most unsatisfactory condition and their diplomas and degrees are . . . of little value." In Hardwicke's opinion, Mexico was a country for serious local and foreign medical students to avoid "as they would the plague."[35] For state officials and medical professionals like Licéaga and Sánchez, Mexico needed to avoid such characterizations if the status of medical education was to improve.

Central to this change was the presence of a new medical curriculum and anatomy museum at the National School of Medicine. Together, they offered a glimmer of hope for state officials seeking to build an army of physicians. Perhaps with the new museum more students would stay in Mexico, while more foreign students would begin choosing Mexico over England, France, or the United States for their medical education. Thus, Sánchez pleaded with President Porfirio Díaz to approve and finance the construction of the anatomy museum, arguing "it would without a doubt, be one of the most beautiful accomplishments of your administration."[36] Appealing to the president's vanity, Sánchez succeeded and received approval for the creation of the museum and his proposed budget on February 19, 1895. Construction began a few weeks later, and Mexican medicine finally seemed poised to reach the level that state officials and physicians like Sánchez believed it was capable of in order to compete in the modern world.

During the construction of the museum, President Díaz's doubled down on his support of the potential the museum could deliver to medical education. By May 1895, two months after he had approved the initial budget for the museum, Díaz decided to increase its annual budget by an additional 3,000 pesos. This took the budget from 5,300 to 8,300 pesos ($1.1 million), which meant the museum could hire additional employees to prepare and conserve even more anatomical collections.[37] The result could be an increase in the ability to attract more students while also showcasing the museum's commitment to shaping Mexican medicine.

But the state's financial commitment to improving the teaching of medicine and medical education—symbolized by the museum—did not stop with Díaz's initial change. Fourteen months later, in July 1896, Díaz had approved another increase to the budget, now 18,000 pesos ($2.35 million per month), which more than doubled the previous year's monthly allowance.[38] Such financial investment demonstrated that the government believed the dissection of cadavers and their subsequent display was an essential component of Mexico's push to improve its medical reputation both inside and outside the country. State leaders accepted that museums were important assets for medical pedagogy because they presented students with more opportunities to study the body than in previous years.[39] As a result, as Professor Sánchez had pointed out in his initial letter, the introduction of an anatomy museum would make "national learning rise from nothing" and "undoubtedly improve" the reputation of Mexican medicine.[40]

The Porfirian state's significant investment in the museum was an essential ingredient for university officials like physician and school director Manuel Carmona y Valle's recipe to improve the curriculum. In a few short years, the museum's collection had already grown to include 1,500 macroscopic and almost 2,000 histological specimens.[41] As a result, Carmona y Valle used the growth of the museum as a tool to advocate for continuous improvements in medical education, particularly by creating a more collaborative environment for students and professors who dedicated themselves to understanding more about the body. Thus, he suggested, the museum and the National School of Medicine could work together to develop courses for students that involved the use of the museum's anatomical collections.

It just so happened that Carmona y Valle had already prepared course descriptions for two classes that he believed would be a perfect fit for the budding relationship: pathological necropsy and pathological histology. Pathological necropsy, a class for third-year students only, would cover "pathological anatomical lesions and their macroscopic and microscopic characteristics," while pathological histology, for fourth-year students, would examine how "disease and sickness" modified organs and tissues.[42] Not only would these courses help establish a strong bond between the museum and the medical school, but they would better prepare students for their exams and careers. Furthermore, in the opinion of influential foreign physicians like Herbert J. Hardwicke, medical universities in parts of the "uncivilized world" with well-trained professors and modern facilities like anatomy museums could "quickly follow in the footsteps of those who have already adopted a high standard of education."[43] This symbiotic relationship between the museum and medical school presented Mexico with an opportunity to improve its international reputation and possibly join the pantheon of modern nations.

Luckily, Carmona y Valle had good friends and supporters in high-ranking positions within the Porfirian state. One such example was Secretary of Justice and Public Instruction Joaquín Baranda, who shared the course proposals with President Porfirio Díaz. Two weeks later, on April 27, 1897, Díaz issued a decree that the National School of Medicine and the Anatomy Museum would work together to begin offering courses on pathological necropsy and pathological histology during the 1897 fall term.[44] In spite of all the new changes underway at the National School of Medicine, some influential physicians still believed Mexican medical education was woefully behind its international counterparts. In a letter to Secretary of Justice and Public Instruction Baranda, Dr. Eduardo Licéaga referenced their discussions from 1893, when they had first tabled the idea of modeling Mexican medical education after European medical universities. In the time that had elapsed since their initial conversations, Licéaga had observed some positive changes occurring in Mexican medical schools. However, there was still more work to accomplish. In particular, he pointed out that since 1893 he had collected detailed information about French and German medical schools to prepare an informative report that would give Mexican state officials something to think about when it came to how best to mold the Mexican medical curricula.[45]

Looking to France and Germany

A medical degree in both French and German universities only took five years, which troubled Licéaga. Students at the University of Paris spent five years completing their studies, while those at the National School of Medicine spent six years. In addition, Licéaga revealed that French students devoted far more time to anatomical dissection than their Mexican peers did despite having less schooling. Outside of assigned class time, professors in France expected their students to take additional courses that did not count toward their graduation requirement but that would complement their medical studies and enhance their understanding of the human body. Government-licensed *repetidores*—individuals who aspired to become medical professors—taught these courses, which included dissection, student operations on cadavers, and how to use laboratory equipment like the microscope.[46] For Licéaga, these expectations were an important reason why French medicine and its students had better international reputations.

Likewise, the University of Leipzig's medical school also took students five years to complete. According to Licéaga, both the curriculum and environment at Leipzig was an even better model than that at Paris and thus should be the example that Mexico would follow. He wrote that teaching in the German university was "extraordinarily rich" with "numerous distinguished and wise professors" and access to "very good" hospitals and a healthy supply of cadavers.[47] For Licéaga, this was a central tenet of what Mexican medicine could become. Mexico too could produce the best physicians, but first the emphasis on dissection had to be placed front and center in the curriculum.

While Licéaga praised French and German medical schools, he did not hold back his criticism for U.S. medical education. Unlike in France or Germany, he wrote, the teaching of medicine in "our Republican neighbor" created unprepared physicians, specialists with "very precise knowledge" that they had acquired quickly during medical school, usually in three to four years. For Licéaga, the American approach to medical education failed to produce knowledgeable physicians. American doctors' short time in medical school had not prepared them to deal with patients with unfamiliar symptoms. This disservice, Licéaga believed, was something that Mexico needed to avoid in order to create versatile and successful physicians: ones who could perform surgery, deliver babies, or assist authorities when it came to autopsying bodies or solving hygienic problems.[48]

To accomplish this goal, Licéaga proposed that Mexico overhaul the country's medical curricula to resemble those of medical giants France and Germany. While the length of medical school would remain six years rather than five, increasing the focus on dissection would produce physicians who knew the body inside and out. They would be ready for any medical problem and thus would be capable of practicing in "the city, town, or even the country."[49] The goal for Mexico—unlike the United States—was to create an army of physicians with a holistic understanding of the body, who would be responsible for improving the health of citizens everywhere and thus guarantee improved public health and the potential for Mexico to find itself alongside France and Germany on the list of the top medical nations. But to institute these changes, Secretary Baranda needed approval from President Díaz. After Baranda sent the president Licéaga's report, which made a persuasive case for increasing the number of courses involving dissection, Díaz responded with a decree on December 15, 1897, that made Licéaga's suggestions a reality.[50] Increasing the opportunities available for students to dissect made students more valuable to Mexico. Furthermore, it guaranteed that unlike in recent years when students left medical school unprepared, students would now be more knowledgeable, capable, and prepared to contribute to improving the health of citizens, which was an essential component of how Porfirian officials sought to manage the population.

Limiting the Theoretical

But for some state officials and medical professionals, these changes were still not enough. For example, Nicolás Ramírez de Arellano, the secretary of the Superior Health Council, believed that in spite of Licéaga's valuable research and numerous suggestions, the curriculum at the National School of Medicine still relied on courses he considered unnecessary. He saw that some professors continued to rely on textbooks for dissection instruction rather than having students practice their dissection skills on cadavers. In a letter dated May 27, 1898, to his friend and National School of Medicine professor Porfirio Parra, the secretary pointed out that propaedeutic (based on patient data) and ophthalmological classes served as impediments to curriculum advancement. What the university needed to do, Ramírez de Arellano wrote, was replace those courses with ones that involved dissection. Furthermore, he suggested that the university abolish a fifth-year course on general pathology and replace it with an entire course dedicated to student dissection on cadavers.[51] Such antiquated methods and approaches to medical instruction supported the negative perception that other medical professionals and organizations had toward the state of Mexican medicine.

Based on course descriptions from the 1899 course catalog, it appears that university officials listened to both Licéaga's and Ramírez de Arellano's advice.[52] In 1899 the university added three new courses dedicated exclusively to teaching dissection: descriptive anatomy, topographical anatomy, and pathological anatomy. Dr. Porfirio Parra, an accomplished writer and professor of physiology who had graduated from the National School of Medicine in 1878, taught the descriptive anatomy course.[53] It met for three hours a week and, according to the course description, sought to emphasize the importance of studying and dissecting cadavers to foster students' knowledge about the body.

Parra also mentioned his specific pedagogical technique in the description that reinforced this approach. His assistant would make the first incisions on the cadaver in front of small group of six to eight students. Next, the assistant would explain orally how to perform systematic dissection before physically removing "the organs in the manner they are discovered." In addition, students would supplement classroom instruction with visits to the anatomy museum to study preserved anatomical parts, illustrations, and engravings that represented the "complicated anatomical layout." All of this would provide students with an educational experience that replaced traditional pedagogical techniques such as lectures and textbooks. According to Parra, students should only use the textbook—written by University of Paris professor Charles Bouchard—as a secondary or tertiary aid. Moreover, Parra posited that if professors insisted on lecturing, they should combine their narrative with an anatomical demonstration using "conserved pieces" or at the very least, "artificial ones."[54] Last, Parra required his students to practice dissection on their own. Students would perform dissections in the university's dissection room in the presence of the professor's assistant, familiarizing themselves with the function and physical characteristics of each organ. Afterward, Parra required them to use their recent experiences to describe orally how to perform a dissection (in order of extraction) and to further explain organ functions as chosen by Parra or his assistant from an "anatomical drawing that is drawn on the blackboard."[55] This method served as an opportunity for Parra to emphasize the important role hands-on activities had in the training of physicians.

Dr. Francisco de P. Chacón—a prominent physician and chair of the surgical anatomy department, whose colleagues upon his death characterized his work as "the history of progress of medicine in Mexico"—taught topographical anatomy similarly.[56] His course focused on students performing dissections on "fresh preparations."[57] As did Parra, he viewed the textbook as a not-so-helpful secondary source for students.[58] Instead, he believed students needed to dissect various cadaver parts on their own, substituting this practical work of "extraordinary preparation" for all written work.[59]

The pathological anatomy course at the National School of Medicine also focused exclusively on tactile techniques. The man responsible for enhancing students' abilities in pathological anatomy was Dr. Manuel Toussaint, a professor of physiology who traveled in the same professional circles as Parra, Carmona y Valle, and Licéaga.[60] Toussaint met with students for four and a half hours a week at Hospital San Andrés. He divided the course into four parts: 1) general technique for autopsies, 2) microscopic techniques and principles of bacteriology, 3) general pathological anatomy, and 4) special pathological anatomy. His course description also revealed his teaching style. For autopsy technique, Toussaint had students watch him or his assistants perform an autopsy, just as Parra and de P. Chacón did. Afterward, students would practice independently, making sure to "pay close attention" to their technique and the color, size, shape, and location of the organs.[61] For students, what enhanced the value of these new courses was the manner in which professors taught the material. Each course focused on presenting students with hands-on experiences, which allowed them to improve their understanding of the human body. With such knowledge, they were better prepared to apply it to living citizens and improve hygiene and health care in Mexico.

The Death of William Scott: The Soldiers of Modernity and an International Incident

In order to enhance hygienic and public health conditions in Mexico, the government turned to medical professionals who shared a common interest—the desire to ameliorate the unsanitary conditions that threatened the health of the city. As medical experts, they claimed special access to the knowledge and skills required to diagnose and offer potential solutions for the city's health problems.[62] In particular, physicians and medical students became invaluable partners for the Porfirian government, since more than any other group they had an intimate knowledge of the human body. They became so invaluable that state officials continued to offer them unwavering support despite international outrage expressed by representatives of the American government over how Mexican physicians and students had handled the death and subsequent autopsy of American railroad workers.

In early February 1900 Powell Clayton, the first U.S. ambassador to Mexico, who would ultimately serve from 1899 to 1905, received a letter from U.S. Consular Agent John H. Farwell. Farwell communicated some disturbing information about a recent incident that had occurred in San Luis Potosí. A former colonial era mining town located 457 kilometers (284 miles) from Mexico City, San Luis Potosí had become a major railroad hub for the Mexican Central Railroad and thus for many foreign railroad employees. But with the growth of the railroad industry in Mexico came increased concerns about employee safety. In particular, the number of American railroad workers who had died in accidents was such that Agent Farwell accused the medical community in San Luis Potosí of having committed routine medical and ethical violations. Farwell accused both physicians and medical students of unnecessarily and repeatedly mutilating the corpses of American railroad workers during their autopsies, an offense that he found warranted some type of investigation on the part of the Mexican government.[63]

The increased presence of American railroad workers had a direct correlation to San Luis Potosí's having become a popular spot during the late nineteenth century for American immigrants who were searching for more independence or an improved standard of living, both of which Porfirian Mexico offered. Cheap land and plenty of employment opportunities in the oil, mining, and railroad industries had attracted Americans to Mexico. In particular, for San Luis Potosí, December 9, 1888, marked the completion of a new international railroad track that connected the United States and Mexico eight miles south of San Luis Potosí.[64] As a writer for the National Association of Railway Agents' monthly magazine *The Railway Agent* explained, "There are Americans in every branch of railroad life in Mexico, the demand for them owing to the rapid increase in mileage, becoming daily much greater."[65] In the late 1880s and early 1890s, many Americans had received support from land development companies who dominated the colonization effort in several cities throughout Mexico, including eastern San Luis Potosí. There, several hundred American colonists—and a handful of absentee American estate owners—claimed most of the land, which covered three districts and was home to more than one hundred thousand Mexicans.[66] Such growth led to increasing tension between Mexican and American citizens over living space and culture, among other things. Thus, Farwell perceived this tension to be further proof that Mexican citizens, including Mexican medical professionals, sought revenge against Americans for taking their jobs and land.

While Farwell's letter concerning the mutilation of American railroad workers disappeared from the historical record for several months, it reappeared on May 26, 1900, when he wrote to the influential Mexican physician Jesús E. Monjarás of San Luis Potosí.[67] In his letter, Farwell pointed out that for the two years he had served as an agent, he was aware of numerous instances in which bodies of American railroad employees who died during railroad accidents had undergone autopsies before Farwell had ever received notification of their deaths. While Farwell was upset about this lack of diplomatic courtesy, he told Monjarás that he took solace in the fact that he believed these autopsies "had been authorized by law."[68] Furthermore, he informed Monjarás that he had told American friends and families concerned about the alarming frequency of autopsies performed on railroad workers, given that most causes of death were quite obvious, to try to follow the rules of the Mexican government.

Nonetheless, the case of William Scott—an American engineer killed on the Mexican Central Railroad—had caused Farwell to question his diplomatic professionalism and acceptance of the status quo. According to Farwell, there was no reason why physicians had to perform an autopsy on Scott's corpse. The train Scott drove had crashed in such a way that the engine had crushed him, killing him instantly. Water and steam from the engine's boiler had also poured out, which had caused Scott's corpse to appear as "nothing more than a bruised shapeless mass of bones with the flesh completely cooked." Thus, Farwell remained convinced that Scott's cause of death was obvious, which rendered an autopsy useless. In spite of Farwell's reservations concerning an autopsy, a local Mexican judge disagreed. The judge issued an order for physician Joaquín L. Delgado to perform the autopsy despite pleas from Scott's friends that "the autopsy could reveal absolutely nothing more than was already known regarding cause of death."[69] But their pleas fell on deaf ears. Furthermore, while Farwell's interpretation of the situation would have been correct under normal circumstances, Mexican law did not require autopsies in cases where the cause of death was "plainly evident." But Scott's death fell into a different category: mysterious deaths. Here, "persons who died of unknown cause or were found in a strange place or in a strange manner" required autopsies in order to determine the specific cause of death, primarily as a way to rule out foul play.[70]

Unsatisfied by the explanation, Farwell used his position as U.S. consular agent to bring his concerns to the governor of the state of San Luis Potosí, Don Blas Escontría. However, unbeknownst to Farwell, the judge who had signed off on the autopsy was actually a federal judge, whom Escontría did not have any influence over, which meant that Escontría was "unable to do anything more than recommend to the judge a more liberal construction of the law."[71] Unsatisfied again by the provided explanation, Farwell decided to bring the issue once again to the attention of his boss, Ambassador Powell Clayton.

Sometime during the month of May in 1900, Farwell, still perturbed, provided Clayton with what he called "eight facts" about how Mexican physicians had performed autopsies in San Luis Potosí. First, he declared that Mexican physicians routinely sawed into the skulls of corpses to extract their brains. At the same time, they also opened the abdomens to examine the internal organs. Second, when physicians conducted autopsies, they "always invited medical students" to witness the procedure, which made sense because the experience was a major part of the curricula shift occurring at medical schools throughout the country. Third, students who attended these autopsies often received multiple opportunities to practice their dissection skills on the deceased. However, their inexperience often resulted in corpses left in a "terribly mutilated state" despite being "patched and sewn" before being returned to their family or relatives. Fourth, in the particular case of William Scott, whoever sawed the skull and extracted the brain had, Farwell wrote, "stitched the scalp inside-out (the inside part of the scalp was placed on the outside) and returned it in this condition to the family." Fifth, Farwell and his staff had "begged and complained repeatedly" to local physicians that the autopsies had been too barbaric. Nevertheless, physicians responsible for them refused to stop autopsying American bodies until they received direct orders from a federal judge.

Despite outrage from various American communities living in Mexico, the Mexican legal and medical networks worked together to protect their medical community. Thus, Farwell saw this relationship as having produced a "mutually protected society that until now, has withstood all attacks." Farwell's sixth fact regarding the poor treatment of Americans, was that he believed the cause of death in Scott's case was clear. Not only had the train engine crushed almost all of Scott's bones, but the escaping vapor and water from the boiler had "burned and cooked" his body. According to Farwell, the result was a cause of death quite visible, which meant there was no need for an autopsy. Lastly, Farwell concluded his report by reiterating that in the case of William Scott—and perhaps the corpses of other American railroad workers—physicians and students had mutilated his body "even more than usual" by sawing, hammering, and chiseling the skull before "throwing the brain to the floor" and executing "other scientific operations" on the corpse.[72] Appalled by the information he had received from Farwell, Ambassador Clayton used his political power to sue the attending physicians who performed the Scott autopsy. His lawsuit forced Mexican officials to require all medical and legal professionals involved in the autopsy of William Scott to complete depositions before having the Mexican Supreme Court issue a verdict on the lawsuit.

The first person to complete his deposition was the judge who ordered the autopsy, Gabriel Aguirre. He argued that cases of railroad accidents involving railroad employee, like William Scott, required further investigation. This often meant conducting autopsies to help uncover the truth when it came to homicides, accidental or intentional. "Some cases are at first view, not attributable to anyone," Aguirre declared, but "afterwards, the autopsy proves the existence of a crime."[73] An autopsy thus provided legal officials with an opportunity to determine if there were more factors involved in the death of William Scott.

Next, the attending physicians who performed the autopsy submitted a lengthy report to defend their actions. Physicians Luis L. Cordero and Joaquín L. Delgado, both of the Civil Hospital of San Luis Potosí, began their report by stating that no physician or student had mutilated William Scott's corpse as Farwell had claimed. In fact, Cordero and Delgado explained, there had never been a case in which anyone had made superfluous incisions beyond what was required for an autopsy. Nor had they or their colleagues used judicial autopsies as an opportunity to teach their medical students more about anatomy, as Farwell had outrageously asserted. In all cases, "without exception," they had proceeded as directed by the judge—and simply carried out autopsies in accordance with the law.[74]

Nevertheless, these physicians also acknowledged that a juridical autopsy was often more invasive than a regular autopsy. They had to not only study the "one or more lesions that explain cause of death" but also had to examine the interior of the skull as well as the abdominal and thoracic cavities. The physicians supported their procedure by referring to an influential legal medicine text from 1877 titled *Compendio de Medicina Legal*. On page 621 of this text, author Luis Hidalgo y Carpio had written that all body cavities required inspection "even when the cadaver has been found mutilated or dismembered" because these acts could have happened after death, as a way to "deflect the attention of the authorities about the true cause of death."[75] Therefore, Cordero and Delgado argued that they had followed proper medical technique and thus had done nothing malicious to the corpse of William Scott during the autopsy.

Their defense went further to prove their innocence. Cordero and Delgado insisted that Farwell's claim that "almost all of the bones in William Scott's body had been fractured" was false, as was the accusation that they had thrown the deceased's brain around the dissection room. As the doctors explained, no cut took place that "did not prescribe to science." They had tried their best to "balance scientific requirements with due respect of the body," which they supported with the inclusion of a detailed narrative outlining how they performed the autopsy.[76] This was important for two reasons. First, the autopsy procedure served as a self-defense tool in the face of Farwell's acerbic accusations.

Second, it revealed the ability of physicians to create an intellectual and linguistic barrier based on specialized knowledge that lay society could, at best, only hope to understand.[77]

The narrative described what they had done to each cavity, starting with the skull. Here, the incision had occurred at the "beginning of the hair line" to the area at the base of the skull, below the bottom of the ears. The physicians then explained how they had lowered the flaps of scalp to cut a line with a saw around the circumference of the skull. From this line, they inserted the "the hook of the hammer" and removed the upper half of the skull so that they could examine the brain. Afterward, the autopsy moved to the thoracic cavity, where—in accordance with the Virchow Method, meaning a Y-shaped incision—the physicians cut the skin in a vertical line down from the "sternum to the pubis bone" and two diagonal cuts toward both shoulders. After removing the tissue and muscle from above the corpse's ribcage, they used special scissors to cut open the ribcage, removing vital organs such as the heart and lungs.[78]

The next step in the autopsy process was an examination of the abdominal cavity and the organs inside it, such as the liver, kidneys, stomach, and intestines. This meant weighing the organs to account for any gains or losses, which could explain a particular sickness along with a brief descriptive narrative of how the organs appeared. At that point, the postmortem examination was over. Attending physicians returned extracted organs to their respective places inside the body and sewed the cavities shut. Then, they—or presumably their medical assistants—"washed the cadaver with a disinfecting solution, dressed him, and delivered the body to the relatives." This entire process took two and a half to three hours to complete, which Cordero and Delgado used as evidence to argue that there was no additional time wherein they could have engaged in frivolous actions, thus rendering the accusations "unnecessary and ignorant."[79] They were confident that state officials understood the reasons behind their meticulous autopsy.

A few days after submitting their description of the Scott autopsy, Delgado and Cordero received support from an influential ally in the Mexican medical profession. Physician Eduardo Licéaga submitted a letter on their behalf, supporting the direction the two had taken in the name of modern science. Licéaga acknowledged Farwell's complaint about doctors performing an autopsy since it was a railroad accident and not a mysterious death. However, Licéaga argued, it was entirely possible that a criminal could have committed a crime of "common order" like homicide, as Luis Hidalgo y Carpio had suggested decades earlier. Moreover, Licéaga explained that the same criminal could use a train accident to "erase his traces and make it appear as if the train was responsible for the death of the victim."[80] According to historian Michael Matthews, most accidents involving trains during the Porfiriato occurred close to towns, villages, fields, and farms where people worked and lived. Many of the victims were railroad employees or innocent bystanders who seemed to be doing routine activities when killed, like traveling to work or walking along the tracks.[81] The carnage that train wrecks brought to people's everyday lives made the tracks a plausible location for murderers looking to dump the bodies of victims. Licéaga also pointed out that according to article 49 of International Law, foreigners employed by Mexican businesses like the Mexican Central Railroad were "Mexican in all that relates to them," which included their sudden deaths. While he understood that members of the American colony in San Luis Potosí "were not pleased that one of their compatriots was taken to the dissecting room at the hospital," Licéaga hoped that they would take solace in the fact that the autopsy was part of the Mexican medical experience and would see how vital it was for ruling out foul play in the death of William Scott.[82] For Licéaga, there was no greater solution to dealing with grief than the truth, which for him meant using the power of medical science.

Concluding his letter of support, Licéaga declared that he believed that physicians Cordero and Delgado had done their job "as it should be—in agreement with the principles of legal medical science." After all, he wrote, even if experts thought they had identified the cause of death, "they should never, under any pretense" settle on their initial conclusions; instead, they should examine "the second or third cavity," for this would provide either additional proof for—or cast doubt on—their earlier findings.[83] Much like he had told students at the National School of Medicine, Licéaga believed that whatever the issue, medical science was an invaluable tool—it outweighed the uneducated and misguided ramblings of uninformed bureaucrats. The truth in the majority of situations, legal or otherwise, existed inside the body. Medical professionals such as Licéaga, Cordero, and Delgado had a professional obligation to uncover the truth, no matter the cost.

Licéaga's opinion on the William Scott autopsy surely influenced the outcome of the case. With all of the testimony recorded and reviewed by Mexican Supreme Court judges, the court issued a verdict in the lawsuit against Luis L. Cordero and Joaquín L. Delgado: the autopsy performed by the physicians was "of legal precision," and "no judge would accept the responsibility of neglecting autopsies." Even if families did not want autopsies performed on their relatives, autopsies were necessary for uncovering the truth and bringing forward individuals who may have committed a crime. Additionally, the court pointed out, physicians "should not avoid performing autopsies" even when faced with "the unlawful measures of foreign consular agents."[84] The future of Porfirian Mexico rested on the unique knowledge that the physician had of the human body. If improving public health was as important as state officials believed it was, then there could not be any interference from amateurs, especially irritated foreign bureaucrats bent on revenge.

The idea that the physician was an essential component of modern Mexico did not just exist within the medical and legal communities. It also appeared in popular newspapers. One such example was the progovernment newspaper, *El Imparcial*, which argued that Mexican medicine based on "proof, data and documentation had replaced eloquence and rhetoric" in society. It went on to argue that physicians were vital for helping state officials analyze public health problems and lending their expertise to curb the uncivilized and immoral behaviors exhibited by the lower classes.[85] The importance elites placed on improving the health of Mexico City inhabitants continued to grow as physicians better understood the threats plaguing the environment and people. In particular, by the final decade of the nineteenth century, the germ theory of disease—first developed in 1883 by German physician Robert Koch—had gained a small following among some members of the medical community in Mexico.[86] But a full-fledged acceptance of this theory would not be achieved until 1903.[87] Even then, miasmatic theory remained a popular explanation for the transmission of disease for both physicians and civilians. For example, at the 1896 American Public Health Association Meeting, Mexican physician Alberto G. Noriega presented a paper on how to reduce miasmatic fevers in the state of Sonora by "planting thick woods around the township," which he argued would reduce the chances that the fever would spread.[88] The explanation that disease spread by microorganisms remained a distant second for physicians and state officials. Medical journals and word of mouth helped to popularize the discovery of new drugs, diseases, or medical theories, especially within the Mexican medical profession.

Bones and Beriberi

Whether physicians accepted bacteriological theory or not, medical students lacked the proper anatomical material required to improve their understanding of both disease and the body. In early 1901 medical students from several different medical schools in the Federal District, including the National School of Medicine, requested the assistance of Governor Ramón Corral to provide them the necessary permission to use the bones of deceased citizens from the ossuaries at Panteón Dolores.[89] Some medical schools were offering students access to bones during class, but it appears as if the students remained unsatisfied with their limited experience. Bones were an indispensable part of medical studies; therefore, if the medical school could not provide them on a regular basis, students had to resort to other measures, both illegal and legal.

But for those students who obtained Corral's permission, he frequently allowed them access to the unclaimed corpses destined for unmarked sixth-class graves at Panteón Dolores. Students disarticulated these corpses in order to build their very own skeletons to aid their studies. Such a case occurred on May 14, 1901, when the director of the School of Homeopathic Medicine requested and received permission to use the body of thirty-three-year-old Francisco Tapia, who had died of pneumonia at the National Homeopathic Hospital the day before.[90] By allowing students to acquire bones from the cemetery or skeletons of the unclaimed poor, the governor contributed to both the medical school and state's desire to advance the understanding of the body. In time, this knowledge could provide the state with medical professionals who could help solve myriad health problems in Mexico City, and eventually the nation. It also demonstrated to the international medical community that the government, under the guidance of President Porfirio Díaz, was truly committed to improving public health and the reputation of Mexican medicine and medical education.

President Díaz and Governor Ramón Corral were not the only influential authorities interested in improving public health and medical education. In May 1901 Professor Manuel Toussaint of the National School of Medicine and National Medical Institute identified a disease—previously undiagnosed in Mexico—that promised to explain many of the capital's (and nation's) health issues. At the National Medical Institute, Toussaint and some of his students had identified why a number of his patients had complained about intense gastrointestinal pain before their sudden deaths. Using the data and observations collected from patient autopsies, Toussaint and the students discovered that the patients had suffered from beriberi, a nutrition-based disorder that affected the nervous system and was caused by vitamin B1 deficiency.

According to Toussaint, beriberi had been unrecognized in all of Mexico until his discovery. The leading reason for contracting the disease was excessive alcohol consumption, which he contributed to the socioeconomic status of his patients, all of whom were from the lower classes.[91] In his opinion, the specific alcohol responsible for the disease was *pulque*—a popular drink among the urban poor derived from the fermented sap of the maguey plant.[92] Toussaint's findings made it easy for state officials to identify potential victims since he wrote that excessive alcohol consumption caused "profound degeneration of the liver," which produced a distinctive gait in those affected. Furthermore, individuals who constantly held their abdomens were more than likely suffering from what Toussaint called a "tenacious intestinal cold," also known as severe and persistent diarrhea, supported by the official death registers which listed diarrhea and enteritis as the leading causes of death in the Federal District in 1878, 1885, and 1903.[93] Autopsies of Toussaint's patients had revealed that besides the liver, the disease also significantly altered other parts of the body—including the kidneys, pericardium, and medulla oblongata.[94] The discovery of beriberi's presence in Mexico City's urban poor prompted Toussaint to have students harvest and preserve affected organs to create a permanent collection for display inside the anatomy museum. The effort to identify and catalog organs affected by beriberi, according to Toussaint, would "benefit humanity and honor the homeland" in a manner never before imagined.[95]

Beriberi's discovery owed much to the power of anatomical dissection, without which beriberi almost certainly would have remained unidentified in Mexico. Nevertheless, its discovery provided state officials with additional proof that regulating the daily lives of the urban poor, especially what they ate or drank, was essential for improving the health of the nation.[96] The desire of Porfirian officials to control the bodies of the urban poor was a hallmark of both the late nineteenth- and early twentieth-century governments throughout the Western world.[97] In the case of Mexico, state officials had tried unsuccessfully to create healthier citizens for more than a century.[98] The most pressing issue had been the poor's overconsumption of alcohol, which prompted officials at the end of the nineteenth century to ban the sale of alcohol, especially pulque, in public places. Government regulations instead required the sale of alcohol to occur inside specialized establishments, hidden from passersby.[99] Limiting the sale of pulque and other alcoholic beverages, state officials, criminologists, and physicians believed, would quickly reduce the number of public intoxication and other related cases and more importantly produce healthier citizens.[100] But the state's attempt to control the sale and consumption of alcohol in a rapidly expanding capital would fail.[101] As the city grew, new pulquerías, cantinas, and cheap restaurants began to appear in the outskirts of the city that continued to keep the alcohol flowing to their customers.

"The Cadaver Is the Best Textbook"

The desire to control and regulate the lives of citizens was an integral part of the Porfirian discourse on how Mexico would become modern. This, however, did not mean everyone was included. Instead, much of this desire provided justification for continual marginalization of the urban poor, widely considered by officials to be the sole reason why the capital's and the country's advance was occurring so sluggishly.[102] Medical science—especially dissection—provided further empirical proof that those most likely to contract diseases such as beriberi were the poor, which made them the scourge of Mexico, according to upper- and middle-class citizens as well as state officials.

The only way for the nation to advance as a whole was for medical education to place continued emphasis on the importance of dissection at the National School of Medicine. In 1902 the description for Dr. Francisco de P. Chacón's topographical anatomy course revealed that in-class instruction was a distant second in terms of popularity to independent student-led dissection. Students only came to the classroom once or twice a week, while they spent the rest of their time practicing dissection on their own or at the university, local hospital, or anatomy museum. Advancing in the medical profession had direct ties to a student's ability to dissect since dissection was paramount for understanding the intricacies of the body. Accordingly, Professor de P. Chacón evaluated students based on their anatomical preparations and continually reminded them, "the cadaver is the best textbook."[103] Many professors at the National School of Medicine agreed with Francisco de P. Chacón's assertion that dissection was the only true method for measuring a student's capabilities for success in medicine.

Reducing the Duration of Medical School Programs

While the cadaver as textbook had become part of the significant changes made to the curriculum at the National School of Medicine, after 1902 there remained room for additional improvements. With a new director in 1906, Dr. Eduardo Licéaga, medical education at the National School would continue to evolve. While Licéaga had been the driving force behind the changes at the university in earlier years, in his opinion, students' practical experiences were still not at the level he desired. Writing in 1906, Licéaga pointed out that he and Secretary of Justice and Public Instruction Joaquín Baranda had worked hard to improve the state of medical education at the school and in Mexico since the early 1890s. For example, he recalled how he had personally overseen multiple revisions to the curriculum to provide Mexican medical students with an instruction that matched the "superior instruction" received by students in European and American medical schools. Furthermore, one of Licéaga's goals had been to create an army of medical professionals who could perform a range of duties—in both rural and urban areas—based on their studies, which included anatomy, physiology, therapeutics, obstetrics, hygiene, and legal medicine.[104]

For Licéaga and many Porfirian state officials, a physician's purpose in the modern world was to be in sync with the desires of government officials. They would be responsible for ensuring the health of the population by instructing them on the values that were most desirable in the modern world, especially as it related to public health. This required a delicate balance between improving the nation's health and performing specific duties related to the monitoring and organization of the population.[105]

However, to accomplish this goal the curriculum at the National School of Medicine still needed further changes. Many physicians still had trouble explaining important concepts such as hygiene or the steps involved with particular surgical operations to their patients. From the perspective of state officials, physicians who were unable to address the importance of hygiene or articulate how they would perform a specific procedure would fail to gain trust and confidence from their patients. These physicians were ineffective agents of the modern state. If the country continued to rely on them, they could damage the reputation of Mexican medicine and the universities that had trained them. According to Licéaga, he and Justo Sierra (the minister of public education) believed that the government could quickly improve the situation by reducing the duration of the medical school program from six to five years, the length it had been during Licéaga's medical school days. Together, they brought this idea to the attention of state officials, who agreed with their assessment.

For Licéaga, shortening the duration of the program provided students with a chance to engage in the practical application of their medical knowledge. It also offered students the chance "to not spend the greater part of their life in preparation."[106] The government saw medical students as effective weapons in the war to modernize the capital and maybe other major cities throughout the country. Doctors could teach their patients practical skills for the modern world that held a particular social value (e.g., hand-washing, use of indoor plumbing, or dietary habits). From there, state officials believed that citizens who exhibited what they considered as uncivilized behaviors would soon change by dressing better, improving their vocabulary, and adopting other traits fitting of the modern Mexican.

Eliminating Theoretical Lessons

Licéaga decided that medical students at the National School of Medicine had trouble explaining hygiene or operating procedures to patients because too many professors continued to rely on textbooks, anatomical illustrations, and other antiquated methods of instruction. Instead, he insisted, the school needed "to completely erase theoretical lessons" and have professors demonstrate "what they teach" via dissection. For him, this meant that each professor needed to focus on offering students tactile learning exercises.

To accomplish this goal, Licéaga proposed that courses such as anatomy and histology require professors to dissect "fresh and conserved pieces" in front of the students. Afterward, students would dissect on their own until "they acquired the knowledge that the professor has given them." Additionally, Licéaga posited that even when it came to improving their understanding of bacteriology—still not accepted as a universal truth throughout Mexico and other parts of Latin America—students should familiarize themselves with the practices outlined by Professor Emile Roux, Louis Pasteur's assistant and head of service at the Pasteur Institute.[107] In 1888 Roux had organized the first regular course on bacteriological technique to appear in medical schools; thus, Licéaga argued, understanding what Roux had taught would provide an important opportunity for Mexican students to be on the cutting edge of medicine.[108] Dr. Licéaga's position as director of the medical school, international recognition as elected president of the American Public Health Association in 1895, and personal friendship with high-ranking state officials, including President Díaz, meant that his suggestions often became a reality.

In fact, his curriculum suggestions also paved the way for the government to provide the university with two new dissection rooms in January 1907. At a total cost of 5,300 pesos ($69,500), the rooms—like the corpse deposits at local cemeteries—combined architectural elements of the past and the present.[109] According to the architectural plans, the walls were made of an indigenous volcanic rock called *tepetate*, while the floors were cement. The room also contained three porcelain washbowls attached to pipes that carried liquids to the sewers, a feature that would create a more hygienic environment for students, workers, and professors.[110] This new equipment, combined with an increased focus on dissection at the National School of Medicine, would facilitate the government's goals of improving public health, modernizing medical knowledge, and creating healthier citizens.

In addition to the new equipment, the focus on transforming the curriculum to a hands-on approach remained ever-present in 1907. For example, the program of study for medical surgeons was just one of many programs at the school that illustrated how indispensable dissection had become in the eyes of university and state officials. The university required first-year students enrolled in this program to take four classes that year, each two hours per week, that focused exclusively on the practice of dissection. This meant that students devoted 40 percent of their classroom hours in a year to anatomical dissection. By their second year, courses containing dissection occupied more than 50 percent of their classroom hours. By the time students had graduated, they had spent more than 1,500 hours dissecting the human body.[111] For the first time in fifteen to twenty years, medical students were more likely to have studied anatomy from cadavers than from textbooks and as a result were more likely to become important players in improving the health of citizens and contributing to the modernization process begun during the Porfirian era.

To celebrate the changes he had witnessed, Licéaga published a short book in 1908 for state and university officials and medical professionals. In it, he applauded "the innovative, new plan" for medical education that now existed at Mexico's preeminent medical school, thanks in large part to his vision for how to change the curriculum. However, rather than thanking state officials and medical colleagues, he made sure to congratulate the Superior Council of Public Education for having made, as he put it, "one of the most important decisions" for medical education and the country's future.[112] The council had mandated that professional exams based on material and questions derived from textbooks no longer had a place in the medical curricula. Instead, medical schools based the majority of their exams on the practical exercise of dissection, as Licéaga had suggested. University professors and state officials, as a result, would soon reap the benefits as students became more aware of the body and thus more able to contribute to improving the health of citizens in Mexico City and, he hoped, in the rest of Mexico.

Even after such change, Licéaga continued to push for more emphasis on dissection in the curriculum. As he put it, students needed to rely on "the direct observation of facts" that the body presented to them, a task that students would not find in books. Instead, examining the layout of organs, studying pathological anatomy, and performing operations on cadavers were all exercises that "would teach their hands" what it meant to be physicians. They would also study "tissues and their modification" by examining slides under a microscope, disarticulating joints and arteries, discussing with professors how to approach hygiene with patients, and spending time in hospitals studying and observing the "natural course of diseases and therapeutic agents" introduced by physicians to their patients.[113] While Licéaga had brought these suggested changes to the attention of the Secretary of Public Education a few years earlier in the fall of 1906, he remained supportive of the direction the school was going in, especially its commitment to dissection and the support the changes were receiving from state officials.[114]

Medicine as the New Religion

The changes made to the curriculum at the National School of Medicine coincided with the fact that for state officials, medicine had replaced religion. Medical students, as a result, had become an integral part of the team responsible for constructing the future. What made medicine especially valuable was the fact that as a discipline it overlapped with other branches of science, such as criminology. Therefore, each discipline provided complementary recipes for how best to improve society based on modern science, which supported the Porfirian government's desire to bring progress and order to the capital and ultimately the country. To this end, state officials considered medical students to be soldiers in the fight for progress, armed not with rifles or machine guns but with a unique understanding of the human body. With such a powerful weapon, students, professors, and state officials all believed that hygienic conditions would improve in the capital as well as throughout Mexico.

A primary belief of this approach was that Mexican physicians could cultivate a "hygienic instinct" in people based on "science, rather than superstition or religion."[115] For Eduardo Licéaga, this meant emphasizing to medical students the idea that they held a unique position in Mexico. During the Porfiriato, Mexican society had given physicians a tremendous amount of trust, which was an attribute that other distinguished professions lacked. Additionally, Licéaga pointed out, the youth of the medical students was an important trait since it represented both honor and purity, which he believed enhanced their righteousness in the eyes of the population. More than any other agent of the Porfirian state, Licéaga believed, the physician would be "both confidant and advisor to his patient," and thus able to exercise particular control over the population.[116] The urban poor were unlikely to visit physicians voluntarily, largely due to cost. Some physicians, however, believed the poor did not visit them because they were superstitious, relying on unlicensed medical practitioners or folk remedies to cure their illnesses. Thus, the job of the physician was to convince weary patients that (modern) biomedicine could change their lives. If the doctor could assure skeptical patients about the benefits of modern medicine, perhaps they would go home and share their positive experiences with their families. The scope of such intervention offered state officials potentially limitless possibilities when it came to managing and transforming society with "a view toward perfecting it."[117] In their minds, the domino effect that could occur from this relationship was an important step for improving health in the capital society and demonstrating a commitment to progress.

The physician's role as confidant and adviser embodied the Porfirian state's desire to turn science, especially medicine, into the new religion of the modern Mexican. In an earlier era, priests were responsible for guiding citizens through the winding maze of religion, providing a list of dos and don'ts. State officials had spent several years challenging priests in the mid-nineteenth century (in the Reform Wars, 1854–67), attacking the powerful position the Roman Catholic Church had held in Mexican society for centuries.[118] However, by the late nineteenth century, science, not religion, offered a prescription for salvation in the modern world.

While Porfirio Díaz loosened some of the restrictions placed on the Roman Catholic Church and its priests by his predecessors, they would never regain the same level of influence they had held in earlier eras.[119] Instead, state officials would designate physicians as true spiritual advisers who could guide the Mexican people through "the science of life."[120] To drive this point home, Licéaga's 1910 year-end address to students at the National School of Medicine mentioned that Mexico had only reached its current "level of civilization" due to the Díaz administration's commitment to science. Thus, in order to continue on the path of improvement, medical students needed to keep working hard so that they could meet "the sublime intentions of the government that so wisely governs us." If students remained committed to medicine, Licéaga reassured them, they would forever be known as "lovers of truth and worthy of trust by Society."[121] By 1910 the Federal District was home to 714 licensed physicians, 23.6 percent of the country's total number of physicians that year.[122] Consequently, every Mexican citizen would turn to the physician as the only source of proper guidance in matters of both health and moral supervision, a powerful and instrumental role that both state officials and Licéaga had envisioned for the modern Mexican physician. Science, not religion, was the prescription needed to improve the health of the country.

Conclusion

Improvements to medical education remained an important feature of the Mexican state well into the 1920s and 1930s, long after Porfirio Díaz fled the country and the Mexican Revolution had displaced thousands and killed thousands more. As historian Rodney D. Anderson has pointed out, President Porfirio

Díaz's fleeing to France in May 1911 "did not mark the death of one era and the birth of another."[123] Instead, change was gradual. Many Porfirian state officials remained in positions of power within state institutions, especially those associated with medicine. At the National School of Medicine and Superior Sanitation Council, officials continued to employ the same techniques and management styles that had existed during the Porfiriato. These officials remained committed to connecting the present with the past as a way to legitimize the role that medicine and public health should play in modern Mexican society.

One way that state officials sought to accomplish this goal was through their pledged support of medical education. Throughout the Revolution, they continued to discuss the role medical education should have in Mexico, despite the social upheaval occurring throughout the country. In 1912 National School of Medicine director Fernando Zárraga wrote a letter about the effects the Revolution was having on the medical school. In his opinion, 1912 "has been abnormal, as it has been for all of the inhabitants of the Republic." Despite a slight reduction in the number of students able to attend classes, "the germ of rebellion" had not dissuaded state and university officials from continuing to view medical education as the solution to many of the capital's (and country's) problems. One such solution was for the school to continue demonstrating its commitment to dissection. On that front, Zárraga proposed that the university install a technology that reflected such commitment: cadaver refrigerators. These were useful because the delivery of fresh corpses from local hospitals appears to have been unpredictable during this era, which meant that anatomical material became even more valuable and needed to be preserved in case of a shortage.[124] Dissection remained the focal point of the curriculum, delivering tangible results for students and thus proving integral for improving the health of citizens.

Inspired by the positive change dissection had brought to the Mexican medical community during the Porfiriato, the university sought to create an official day that celebrated them. So on July 31, 1918, the school created El Día de Cadáver Anónimo (the Day of the Anonymous Cadaver). According to the medical school's director, students, professors, and staff all had a "moral obligation" to demonstrate their "gratitude toward the souls of the cadavers whose names are ignored."[125] The day itself, according to newspaper accounts and university records, included students and professors performing musical pieces such as Charles Gonoud's "Je Veux Vivre (Juliet's Waltz)"" and Puccini's "Vissi d'arte" (I Lived for My Art). Additionally, medical students recited original poetry about what the cadavers had meant to them. The celebration even included an appearance by sitting Mexican president Venustiano Carranza.[126] For school officials, the most important thing students and professors needed to remember was that these cadavers had contributed to their education and were instrumental in the improvement of public health in the capital and the country, which would make individuals like Dr. Eduardo Licéaga proud of his alma mater.

The Day of the Anonymous Cadaver also reminded students, professors, and attendees that the dissections performed daily at the university played an important role in shaping the modern culture of the capital. Too often, sometimes captured by amateur and professional photographers, students had treated corpses less like humans and more like toys, which had led "students to go too far, committing true sacrilege, for only a joke." At some point, for example, students at the National School of Medicine had placed the severed bloody hand of a cadaver in a nearby park for passersby to find.[127] Such behavior lacked professionalism and appreciation for the dead. To demonstrate the school's commitment to civility and progress, university and state officials wanted the celebration to exude respect for the dead. At the very least, students could put aside their immaturity for one day in order to display a modicum of respect for the corpses they dissected daily, which were the foundation for their improved knowledge and medical careers.

The commitment to dissection remained so instrumental that from the 1920s to the 1950s, it created a heated rivalry between the

National School of Medicine's medical and dentistry schools. In 1920 the Secretary of Dentistry Department wrote a letter to the director of the medical school to ask if he could use "his valuable contacts" to help dentistry students acquire cadavers from a local hospital.[128] This request sparked a thirty-year competition about which students deserved to receive cadavers. State institutions such as poorhouses or mental institutions (for example, La Castañeda) routinely denied the dentistry department access to fresh bodies. However, they continued to provide the medical school with corpses for all of its courses that involved dissection, as they had done for years.[129] Such commitment to dissection highlighted the fact that regardless of who was in charge of the government, access to cadavers remained a valuable weapon in medical education, even among colleagues. The fight to improve public health and bring lasting modernity to Mexico hinged on whether or not students had access to the freshest anatomical material, and the medical school could continue to say it had such access.

3

Wet or Dry Remains

Funerary Technology and Protecting Public Health

Mexico City's attempts to change its reputation from a dirty and backward capital to a shining example of modernity in the Americas was not limited to just new forms of transportation for the dead or a renewed focus on dissection in medical schools. In particular, Mexican state officials' desire to protect public health and remove rotting corpses from the capital's streets helped bring into existence new business opportunities for many entrepreneurs inside and outside of Mexico. Funerary technology, as officials categorized it, offered both amateur and professional inventors—many of whom thought of themselves as scientists—two opportunities: a chance to help the Mexican government improve public health and a chance to make a fortune.

In mid- to late nineteenth-century Mexico, the number of registered patents was far less than in the United States. One of the main reasons for this was the fact that, as historian Edward Beatty has pointed out, Mexico was a country with "relatively low levels of human capital and relatively backward state of existing productive technology." But all of this began to change once President Porfirio Díaz repeatedly *won* reelection. He and his state officials saw the lack of technology as an opportunity to implement policies that recognized patents as a vital source of technology transfer that would help keep Mexico on the path to progress.[1] Conditions in Mexico after 1870 led the country to import its technology because it had failed to adopt the technologies of the first industrial revolution, which, as Edward Beatty has pointed out, resulted in Mexico not having "the particular kinds of human capital necessary to assimilate the knowledge and expertise embodied in technology imports."[2] Entrepreneurs and state officials had to rely on importing both technology and expertise from abroad. For investors interested in building potential wealth, many of whom were foreigners, Mexico presented a unique opportunity to make significant financial gains. To make his country more appealing and strengthen the Mexican economy, President Díaz found ways "to make selective commitments to privileged groups of asset holders" and present Mexico as a country rife with opportunities for untold fame and fortune.[3] Included among these groups were individuals from around the world who submitted patents for new technologies developed exclusively to improve public health and hygiene in matters related to death and dead bodies.

While the Porfirian state struggled with creating a modern capital that was safe, healthy, and organized, the twentieth century ushered in innovative funerary technologies that sought to assuage the health problems found in Mexico City. In the last two decades of the nineteenth century, there were only six patents submitted that proffered methods of handling bodies. Of those six, five dealt with coffin construction and one mentioned a method for conserving corpses.[4] However, in the first decade of the twentieth century alone, there were sixteen patents that related to bodily disposition methods, which represented a 267 percent increase in ten years. These funerary patents appealed to state officials and elite and middle-class citizens because they reinforced their values and sensibilities, which centered on improving public health in a city where a growing number of dead bodies could be found scattered in the streets. These new patents presented state and medical officials with an opportunity to use new forms of technology to protect public health and foster a hygienic environment that demonstrated the city was capable of becoming modern.

While several studies have focused on the city's environmental problems, such as low-lying areas prone to flooding that led to elites' creation of their own new neighborhoods, none have addressed how state officials planned to deal with the growing problem of dead bodies that threatened the city's public health conditions.[5] This chapter argues how the Porfirian state's desire to improve such poor public health helped introduce technology that focused exclusively on protecting both living and deceased citizens from any health threats posed by decomposing corpses. Funerary technology promised to slow the decomposition process, but ultimately it had the ironic effect of accelerating decomposition in many instances. These technologies also helped to reinforce existing class divisions by offering five methods of hygienic disposition that would contribute to the modernization of the capital and hopefully the country: the coffin, the burial vault, topical embalming, arterial embalming, and cremation. While upper- and middle-class citizens were the only groups that could afford this technology, state officials still believed that exposure to such methods would provide the lower classes with an opportunity to learn how to deal with the dead in a modern way.

Brief Background of Patents in Mexico

State officials in Mexico, like those in the United States and Europe, believed that domestic patents were an important part of the technological and industrial development of a modern country.[6] Furthermore, whether accurate or not, Mexican state officials were quick to promote, especially to foreigners, the idea that inventors maintained rights to their inventions, which was an attractive feature. At the same time, the government encouraged Mexican inventors to submit patents often because if granted the inventions had only a limited lifespan before coming into public domain. Once expired, the patent information—both the narrative and accompanying drawings—became available to anyone who wanted it, which could have financial and intellectual consequences as others might look to improve on the original.

Despite these opportunities, Mexican inventors were not the

source of new funerary technology. Instead, much of it came from foreigners, especially citizens of the United States, who viewed the Mexican market as enticing because of its low application costs and more importantly the perception that it was a land rich with economic opportunities. For many foreigners, Mexico was also attractive because the government had doubled the length of patent protection in 1890 from 10 to 20 years. It would also soon reduce application costs substantially after 1903, from 50 pesos ($1,440) to 5 pesos ($144).[7] In terms of patent length, inventors who paid a percentage of the total fee (5 pesos) received *provisional* status for their invention, guaranteeing their patents for one year. However, if they wanted to maintain rights for longer than one year, the government required an additional payment of 35 pesos ($1,010). Then patents were granted *definitiva* status and guaranteeing exclusive rights and potential royalties for an additional 19 years.[8] The Mexican government modeled these patent rights on those found in the United States, focusing on "investment in invention over investment in innovation" and strengthening the rights of foreign inventors by giving them equal protection and priority rights.[9] This confluence of events made the Mexican market extremely affordable and attractive to inventors outside of Mexico with significant disposable incomes since their inventions had the potential to yield a fortune.

In comparison, for patent applications submitted in the United States, the cost after 1861 was $35 ($2,740).[10] While the cost may seem high, the average annual salary for all laborers in the late nineteenth-century United States was $437.96 ($60,400), and most of the Americans who submitted patents worked in industries that paid substantially more.[11] In Mexico, however, the average salary was far less, with the daily wage of many laborers in the early twentieth century estimated at one peso ($68).[12] Thus, the United States remained a less affordable option for many inventors, especially Mexicans, during the late nineteenth and early twentieth centuries.

Yet access to protection and rights appears to have mattered less than the initial cost of securing patents to foreign patent-seekers. Patenting fees in Mexico, while low compared with those found in the United States, still represented a tremendous financial burden for Mexican inventors. Fees ranged between two and three times the average annual per-capita income, even with the fee reduction in 1903, which meant that patenting was an expensive venture for most Mexican citizens. Such uneven income distribution, a hallmark of Porfirian Mexico, along with the "lack of effective capital markets" for those without political connections, resulted in the majority of funerary patents coming from foreigners who had the finances to cover the fees and who also believed they could simultaneously improve their personal wealth and solve Mexico's corpse dilemma.

Fashionable Coffins

One of the more traditional patents submitted in this period was the coffin. This method of bodily disposition was one of the oldest; it had grown out of livery stables around the world because they supplied the transportation method for the deceased (horse and cart). But early coffins were often crude and offered little protection for public health. Nevertheless, they remained a popular disposition method for elite members of society, who believed that burying a body without a coffin was something that only the poor did, not those with money.

For the well-to-do, the coffin had become a symbol of class difference, a way to differentiate the elite deceased from ordinary dead. The popular and intricate designs found among coffins in Mexico in the early twentieth century mirror those found in the United States in the nineteenth century. Around the year 1800 U.S. citizens began investing in stylish coffins to highlight the deceased's status for those who attended the funeral and for those waiting in the afterlife. The practice appears to have had its origins with British colonists, who used "simple, unadorned wood receptacles" for those of ordinary backgrounds, while aristocrats had coffins made of mahogany, adorned with expensive fixtures and jewelry, and lined with silk.[13] The deceased from well-to-do backgrounds often remained unburied for days while friends and relatives paid their respects, so the design of the coffin, for these types of individuals, had to match the regal furnishings that had surrounded them in life. It was part of an emerging impulse in U.S. and Mexican society to create a physical symbol of beauty that matched the socioeconomic status of the dead. As a result, this helped foster an environment that sought to commodify death.

Subsequently, this led to the growth of the professional funeral industry in both countries. It would become such a burgeoning business that by the turn of the twentieth century, North American funeral professionals like W. P. Hohenschuh advised colleagues "to change the style of caskets, trimmings, linings, and robes as often as possible" to take advantage of the emerging consumer culture surrounding death.[14] Nevertheless, the shape of the majority of coffins remained basically unchanged; coffins still conformed to the shape of the human body, tapered from the shoulders to the head and from the shoulders to the feet.[15] For example, one of the most popular designs in the United States was the Fisk Metallic Coffin, patented in 1848 by Almond Fisk. His tailored coffin reduced the amount of material used, which meant that it was comparatively lightweight despite being made of metal. The coffin itself resembled a sarcophagus, with an engraving of the arms folded across the chest and a cross positioned between the hands. Meanwhile, the head of the coffin resembled an underwater diving helmet with a glass plate that allowed the deceased's face to be visible. Fisk claimed that not only could he alter his metal coffin to accommodate any size person but that his coffin also created a hermetic seal to slow decomposition of the body. It even included an optional feature that allowed families to fill the coffin with "any gas or fluid having the property of preventing putrefaction" in order to preserve the body in a lifelike appearance, a feature whose popularity would skyrocket by the early twentieth century, especially in Mexico.[16]

Fisk enjoyed the immediate benefits of his coffin's popularity. The additional money led him to offer more coffin accessories that would appeal to a variety of consumer tastes. But this was short lived, as he sold the manufacturing rights to the cof-

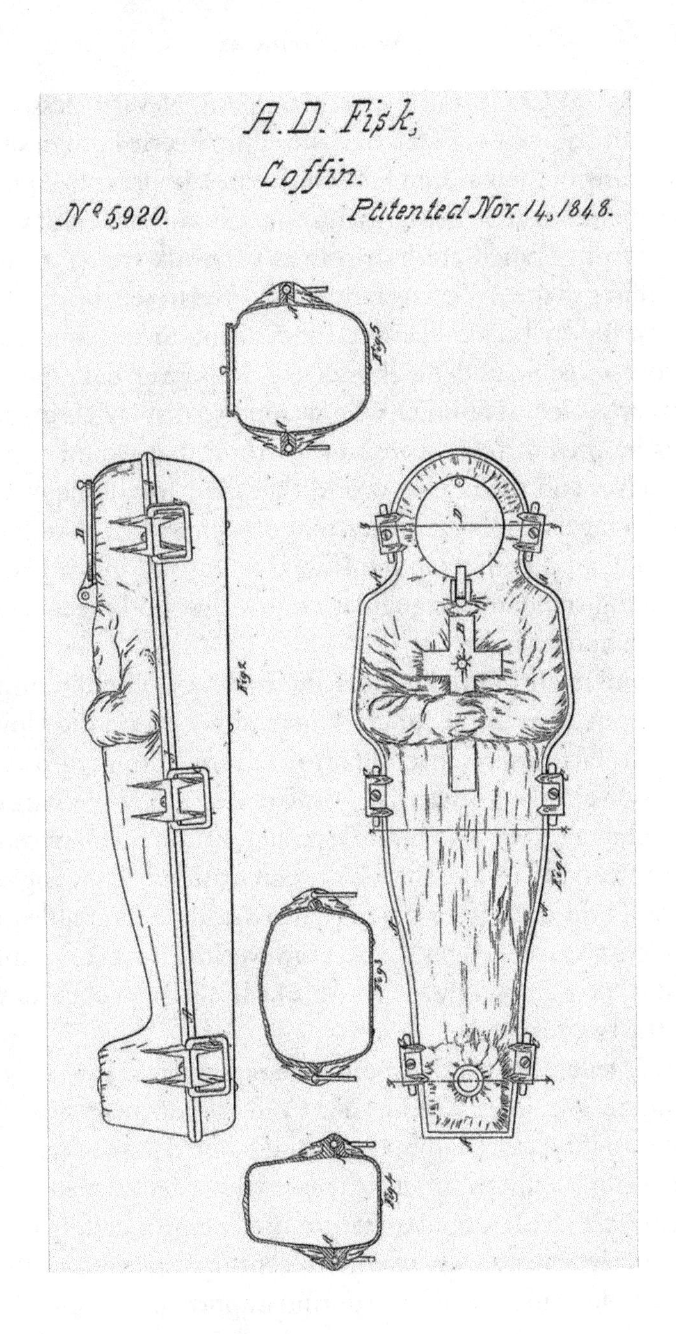

FIG. 5. Fisk Metallic Burial Case, 1848, patent sketch. U.S. Patent Office, A. D. Fisk, inventor, patent 5,920, issued on November 14, 1848.

fin in 1853 to Crane and Barnes Company. Nevertheless, both companies quickly realized that the coffin's eerie human shape irked many of their customers' sensibilities. In order to comfort them, manufacturers began to introduce decorative cloth wraps "made from French cloth, trimmed with silk fringe" to hide the coffin's shape. Yet many families desired to see the deceased one last time, which influenced the decision to keep the face of the corpse visible under a glass plate.[17] Moreover, the expensive wraps appealed to families who wanted to display their social status by providing a disposition method that would remind themselves and the outside world that their loved one was far from average. By 1860, however, manufacturers had abandoned the Fisk Coffin entirely in the United States and instead began producing standard rectangular coffins that no longer resembled the human figure.

Despite its short-lived popularity, the Fisk Coffin did inspire the designs of amateur funerary inventors outside the United States. In fact, several inventors from Spain submitted patents in Mexico clearly modeled in a similar vein to Fisk's coffin. On November 22, 1907, J. Ramón Díaz, a citizen of Spain, received definitiva status in Mexico for his patent titled "Drawing of a Special Form of Lids for Coffins," which closely resembled the Fisk Coffin's sarcophagus shape. However, the lid, Díaz pointed out, was "new and original" (see fig. 6).[18] But Díaz's coffin lid was only the beginning.

Díaz's fellow Spaniard Joaquín D. Tames received patent rights on May 22, 1910, for his "Model of a Coffin or Mortuary Box" that was rectangular, carved from wood, and tailored to the deceased's dimensions.[19] Tames's coffin also came with a crystal plate above the corpse's face, a popular feature of expensive coffins of the late nineteenth and early twentieth century, especially of those used for Mexican state officials during important state funerals.[20] For fashion-conscious families, his design also offered additional accessories such as an option for "decorating the interior or exterior with any fabric desired." Additionally, the coffin's design and fashionable approach to death included specific information

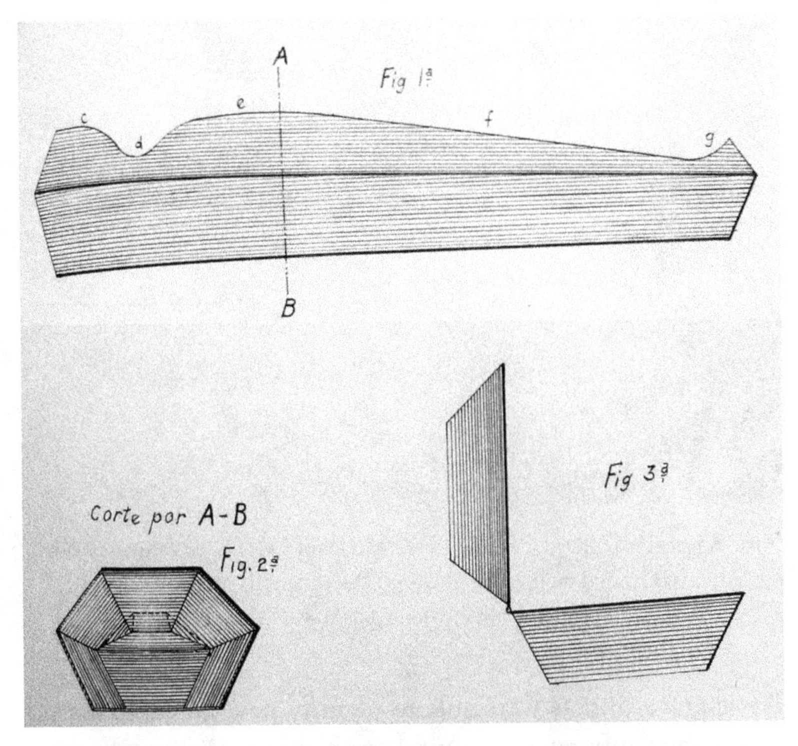

Fig. 6. J. Ramón Díaz, patent 4,150, "Drawing of a Special Form of Lids for Coffins." Archivo General de la Nación, Grupo Documental-Patentes y Marcas, legajo 307, expediente 84, November 22, 1907.

about the coffin's safety features, which included four bolts in the lid that would ensure the coffin remained sealed from potential nuisances such as animals or graverobbers.[21] This was a specific feature that gained traction in the early twentieth century since it offered a certain level of protection for public health.

In many instances, inventors included bolts and gaskets as part of a coffin's design in order to showcase its protective features to families concerned with potential graverobbers and to state officials concerned with public health. Throughout the United States and Europe during the nineteenth century, protecting a loved one's coffin from unwanted entry was a primary concern as medical schools paid scurrilous individuals, popularly known as body snatchers, to steal newly buried corpses from cemeteries.

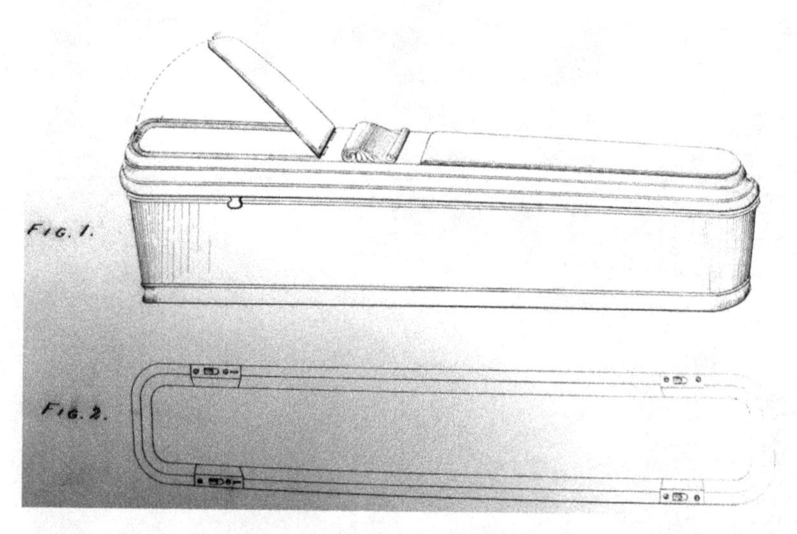

Fig. 7. Joaquín D. Tames, patent 4,306, "Model of a Coffin or Mortuary Box." Archivo General de la Nación, Grupo Documental-Patentes y Marcas, legajo 307, expediente 86, May 28, 1910.

These graverobbers were able to identify new graves based on the presence of new dirt around grave sites, which meant fresh anatomical material they could deliver to students for a higher price. Yet the evidence for Mexico suggests that body-snatching was nonexistent; perhaps this was due to the religious significance the body held in Catholic Mexico as well as the fact medical schools had a plethora of unclaimed corpses to choose from through local hospitals or poorhouses.[22] Moreover, while these bolts and gaskets did protect the corpse from potential invaders, they also served an ulterior purpose.

During the late nineteenth and early twentieth century, Mexican state officials placed great emphasis on protecting public health from noxious gases and odors emitted by decomposing corpses found in the street. This obsession was so prevalent in the scientific community that inventors tailored their patent applications to reflect the degree to which their unique patent could protect public health from diseases that many believed corpses to emit. Despite the fact German medical officer Robert Koch had developed germ theory in 1883, many leading Mexican, American, and British scientists still believed that miasmas—infectious mists or vapors emanating from swamps, garbage, raw sewage, animal carcasses, or decomposing bodies—were responsible for how diseases spread.[23] This theory was so prevalent in nineteenth-century Mexico that state officials had cited it as the reason why burials had to move away from beneath church floors to suburban cemeteries.[24] For many Mexicans, including people within the medical community, miasmas as a source of disease made the most sense since places that contained dead or dying people like cemeteries, corpse deposits, prisons, or hospitals were not popular places to live.

By the early twentieth century, miasmas remained the common agent blamed for spreading diseases in Mexico. Thus, secure coffins reassured state officials that the potential disease found in decomposing corpses would not affect the public. This emphasis on protecting the health of city residents presented inventors with a unique opportunity to profit from the culture of fear surrounding miasmas. In addition to offering protection, hermetic coffins also needed to be fashionable. Inventors had seen the success sculptors had experienced, both financially and socially, from the desires of elite families to highlight their wealth and cement their status by erecting monuments or constructing headstones and mausoleums made from bronze, stone, or marble.[25] If funerary professionals adopted new coffins, they could offer families what they had sought when they hired sculptors to design elaborate tombs or lifelike busts to honor the dead and console the grieving: a physical symbol of both status and significance to showcase to the living world. Furthermore, coffins were part of the Porfirian obsession over technology that state officials believed would help revitalize the capital. While the lower classes may not have used coffins for various reasons, most notably cost, state officials believed that repeated exposure to modern coffins (as well as other funerary items) would provide the lower classes with a model that they could follow to become more productive citizens.

Hermetic Coffins

While the design and potential accessories for the traditional coffin during the nineteenth century had improved the capability to preserve the corpse, they failed to protect public health. State officials and residents still believed that slowing the rate of decomposition would provide them with a healthier environment, free from disease and potential contagion that existed inside putrefying corpses. To offset the poor sanitary features of traditional coffins, the Porfirian government granted eight patents—five for hermetic coffins and three for impermeable burial vaults—to patentees whose applications and designs promised to reduce the decomposition rate, thus protecting citizens from diseases and sicknesses associated with decomposing bodies.

An American inventor from Warren, Pennsylvania, named John Charles Fremont McGriff submitted the first of these patents, dated October 1, 1903. He had already received patent rights for a similar coffin in the United States in 1901.[26] Nevertheless, in Mexico he paid for the twenty-year life of his patent upfront and as a result received definitiva status for his "Improvements in Coffins."[27] The first of the improvements he promised was for his coffin to be "absolutely sanitary." Additionally, his coffin also provided state officials with another feature that appealed to them: a method for protecting public health. While the corpse rested inside the metal cylinder, at the feet of the deceased, there was a partially sealed tube that presented families with an opportunity to "introduce a preservative" in gas form. The corpse would absorb the gas preservative, which would reduce decomposition and thus limit the spread of disease. Fremont McGriff assured state officials that the gases and liquids associated with decomposition would not escape the coffin and enter the atmosphere but would stay inside the coffin. This was an attractive feature for state officials who worried about whether or not corpses could spread disease throughout the city and infect living citizens.

The second of Fremont McGriff's improvements reflected how the commodification of death was slowly becoming an

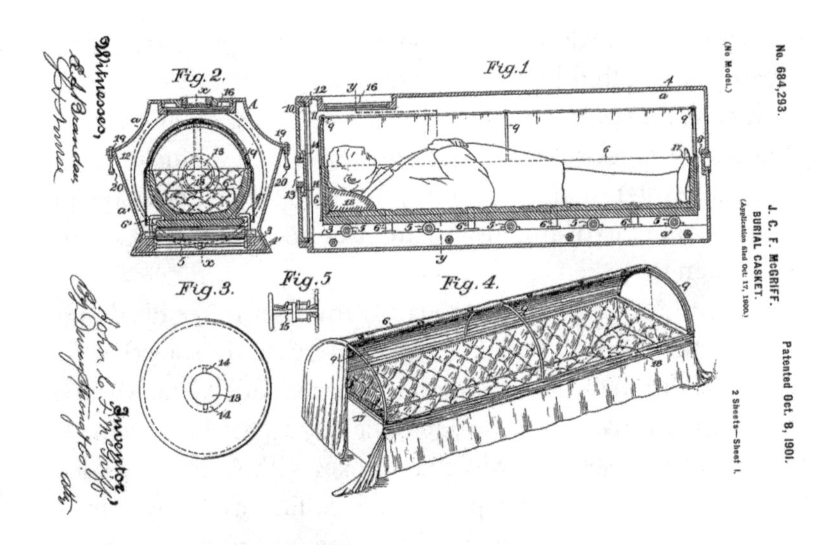

Fig. 8. John C. Fremont McGriff, Mexico patent 3,306, "Improvements in Coffins." There is a small hole to the right of the deceased's feet where a gas preservative could be introduced. U.S. Patent Office, J. C. F. McGriff, inventor, patent 684,293, filed on October 17, 1900, and patented on October 8, 1901.

expected feature for all funerary items, especially coffins. In addition to the available sanitary features, the coffin also provided the family with one last view into the coffin through a small window "made from glass or crystal" above the face of the deceased. While the standard model came unadorned on the inside, Fremont McGriff offered his clients a chance to demonstrate their social status by adding additional upholstery of their choosing. This included the option to line the coffin with "porcelain or some other indestructible substance" that was expensive, which meant only a select few could afford the luxury.[28] Additionally, Fremont McGriff's coffin also contained detachable head- and footrests, as well as a sliding curtain "to cover the body or expose any part to view, if desired."[29] The inclusion of a sliding curtain was inextricably tied to notions of elite identity and "the definition of the body in which caution, modesty, and modernization would dominate."[30] Knowing the deceased was resting comfortably soothed the conscience of relatives and allowed the family to preserve the deceased's modesty. Unlike the corpses of the urban poor—often buried nude without coffins in unmarked sixth-class graves in large, public cemeteries such as Panteón Dolores—Fremont McGriff's coffin was for an exclusive clientele. Those interested in his coffin were wealthy elites who continually sought to distinguish themselves from the rest of society, even when they died.

Fremont McGriff was not the only inventor to see his design as a way to showcase social status, modesty, and protection for public health. Arcadio Hernández, from Guadalajara, Mexico, received definitiva status on October 1, 1903, for his patent, "A Construction System for Mortuary Boxes with Automatic Closing." From an aesthetic perspective, the coffin was different from McGriff's because of its distinctive shape, as it resembled a life-size jewelry box. The standard model included crystal or glass rectangular pieces on both sides of the coffin as well as near the corpse's face that permitted family members to see inside. If the family wanted more modesty for the corpse, there was the optional installation of fine accessories, like curtains along the sides of the coffin.[31] Families could draw them closed for privacy or leave them open, as one last symbol for the living world to see that the opulent surroundings the deceased had enjoyed while alive would continue. While this was similar to Fremont McGriff's coffin, the most important similarity was how Hernández's sought to protect public health. A several-inch-deep canal around the top of the coffin with thick rubber molding provided a hermetic seal when closed and thus reduced the threat to the living posed by the putrefying body.

Moreover, the most innovative feature of the coffin was its double-bottom construction. While difficult to spot in the patent sketch, the application's narrative detailed how the bottom of the coffin contained several large holes covered by swivel caps. At the moment of burial, cemetery workers removed the caps, which allowed liquids from the body to drain beneath the coffin and into a second bottom section that functioned like a drain pan, three to four centimeters below the coffin. This feature would ensure that rather than liquid collecting in a pool beneath the corpse, creating a sticky, unhealthy residue, the liquid would drain out of the coffin.[32] This made Hernández's design particularly attractive to state officials. The liquids leaving the decomposing body collected in the drain pan, which trapped them and thus reduced the threat decomposing corpses posed to the city's public health.

While Fremont McGriff, Hernández, and Porfirian state officials all believed that these coffins had hermetic properties, the reality was much different. Neither coffin actually delayed decomposition. Pathologists now understand that anaerobic bacteria thrive in airless atmospheres. Their understanding of bacteria was still in the embryonic stage, as miasmatic theory remained a popular explanation for disease, even among trained medical personnel.[33] Instead of protecting the body, hermetically sealed coffins accelerated decomposition, a circumstance about which government officials and physicians were most likely unaware.

Indeed, the prevalence of miasmatic theory was perhaps responsible for the introduction of patent number 3681, which combined hermetic properties with chemical preservatives to reduce decomposition rates and protect public health. On April 26, 1904, Mexican citizen Carlos Navarro Mora received definitiva status for what he called "The Automatic Closing Mortuary Box." To reduce the public's exposure to potential threats posed by decomposing corpses, Navarro Mora used a dual coffin design. The coffin containing the body was made of metal and used a rubber seal along with a series of hand-turned locks to prevent air or fluids from entering or leaving. The metal coffin fit inside a larger wooden coffin with the same seal and locks, which created a reinforced barrier against escaping gases. If any did escape, they would have to pass through two layers before reaching the air that city residents were breathing.

In addition to multiple layers to reduce possible contamination, the coffin contained two preservative cushions—one from the upper back to the feet and the other under the head. He filled each cushion with a liquid he described as "general alcohol" that he guaranteed eliminated "deleterious miasmas" exiting the decomposing corpse.[34] The cushions were more than just decorative; the corpse's skin absorbed the alcohol and then would exhibit characteristics similar to partial topical embalmment. This was an approach similar to what British sailors had used to transport the corpses of sailors who died at sea—including that of British Naval war hero Admiral Horatio, Lord Nelson, whose pickled body arrived in England in 1805 stuffed into a barrel of rum.[35] While Navarro Mora's coffin offered state officials a simple approach to protecting public health from decomposing corpses, his design did not foster radical change within the Mexican funeral industry. Nonetheless, his coffin and preservative cushions inspired other inventors to adopt some form of topical embalming for corpses.

While many of these inventors have slipped into obscurity, leaving almost no details of their lives behind, insights can be gleaned from the life of one inventor whose story has survived. American inventor and physician Monroe S. Leech received definitiva status for what he called "Improvements in Mortuary Boxes" on March 1, 1907.[36] Born October 14, 1845, in Shelby, Ohio, Leech served in the Union Army in the 163rd Ohio Regiment from May 1, 1864, until the end of the Civil War. After the war, he studied medicine at Western Reserve College (now Case Western) in Cleveland; despite his claims of having graduated in 1867, he did not graduate.[37] Nevertheless, he opened his own medical practice in a small town outside Kansas City, Missouri, where he stayed for three years before returning to Ohio to enroll at the Eclectic Medical School in Cincinnati.[38]

According to the scant records of the Eclectic Medical School, Leech graduated in the winter 1870–71 term, having paid for two courses and completing a thesis on congestive intermittent fever (a severe cold in patients whose health was already deteriorated).[39] After graduation Leech returned to Missouri, where he practiced medicine for another ten years. In 1881, citing health problems, he moved to Chicago, where he took more classes and graduated once more, this time from Rush Medical College in

1882.[40] Leech remained in Chicago until his death on June 24, 1909, from complications associated with nephritis, a disease caused by "the abnormal production or accumulation of acid in the cells of the kidney."[41] His sudden death also received coverage in several newspapers, largely due to the death of his pet monkey, which reportedly "starved itself to death" after Leech died.[42]

Like other American recipients of Mexican patents in this era, Leech had also received a patent for his mortuary boxes in the United States a year earlier on January 9, 1906 (patent number 809,573).[43] He employed a traditional coffin design but added one new feature: a glass dome over the upper half of the coffin. But Leech's invention was not entirely new, as it closely resembled in design one submitted in the United States in 1859 by George W. Scollay of Saint Louis.[44] Leech's glass dome, however, provided a constant hermetic seal because Leech sanded the coffin's edges and bound it to the glass using "cement or other binding material."[45] Additionally, he attached a machine to extract the oxygen from inside the coffin and replace it with disinfecting gas. This patent appealed to both Mexican state officials and affluent citizens since it combined two of the most important features for early twentieth-century coffins: protection for public health and preservation of the deceased's appearance.

The glass-domed casket failed to gain a following in Mexico. Its design, however, did appear to usher in a distinct change in how inventors approached the construction of new coffins. The funerary patents that followed Leech's began to incorporate cement as a primary building material. Cement, as Leech had explained, offered inventors hygienic properties that far exceeded those found in traditional coffin building materials.[46] One such example appeared on September 5, 1911, when American Elijah D. McDonald of Los Angeles, California, received definitiva patent status for his "Mortuary Box," which he described as an "indestructible and waterproof coffin."[47]

Elijah McDonald was no stranger to the patenting process. Before submitting his "Mortuary Box" plan in Mexico, he received several patents in the United States for a trolley, a self-lubricating

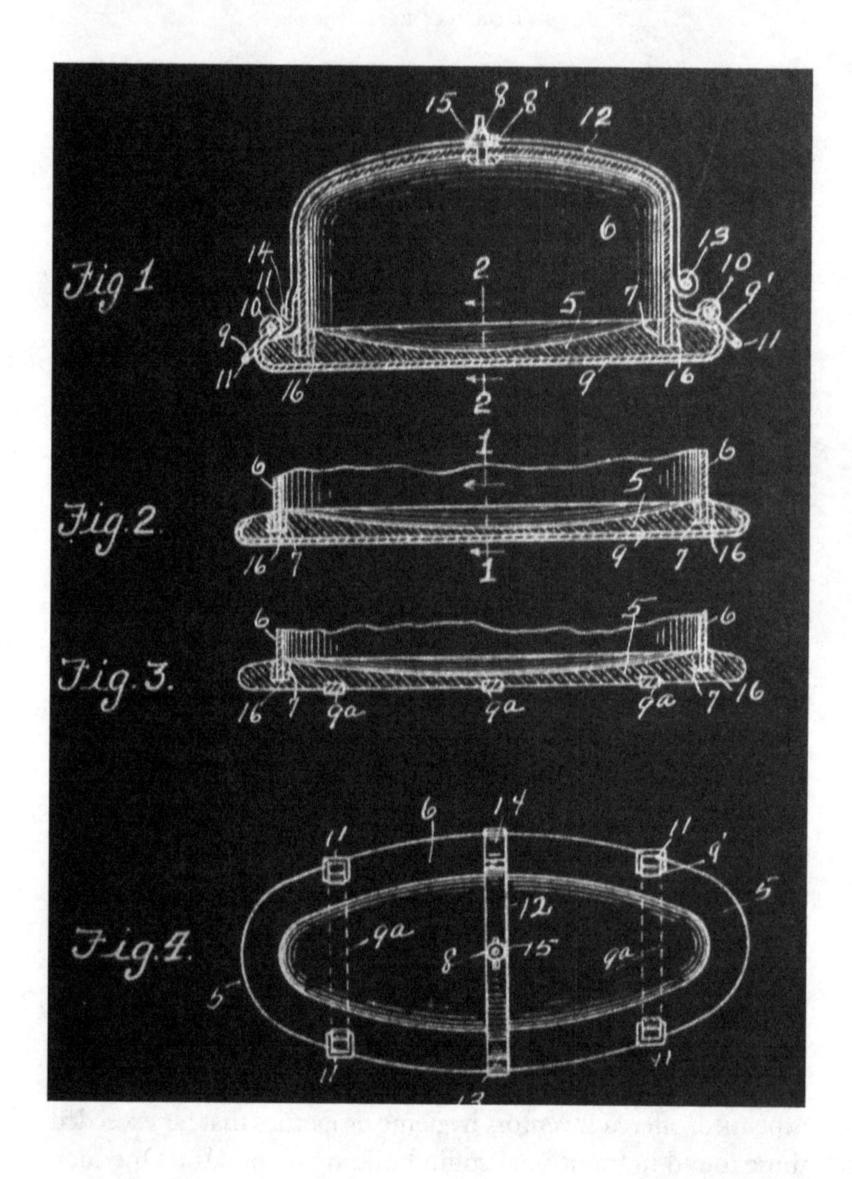

FIG. 9. Monroe S. Leech, patent 6,620, "Mortuary Box." The glass dome is listed as figure 1. Archivo General de la Nación, Grupo Documental-Patentes y Marcas, legajo 307, expediente 80, March 1, 1907.

wheel, a new form of railway construction, and reinforced concrete.[48] According to his patent application, the most innovative feature of the coffin was its great strength "in proportion to its weight," a feature that he claimed was absent from other coffins. His coffin contained a basket made from a single piece of woven wire fabric, which was molded over the frame of the casket. The result, he explained, was a coffin "strong at any given point," no longer vulnerable to people, animals, or dirt. McDonald claimed that despite the impenetrable nature of the coffin—capable of withstanding a pressure of 1,100 pounds per square inch—it weighed the same as a coffin made from solid wood. In addition to being lightweight, McDonald made sure to point out, his coffin was also extremely hygienic. To achieve this, he covered the inside of the coffin with a special solution made from "one-part Portland cement, one-part gypsum, one-eighth part magnesium, one-part molasses, and one-part water," which he guaranteed made the coffin impermeable to man and nature, features that appealed to both state officials and elite families.[49]

Two of McDonald's ingredients had rich histories in the burial process. For example, gypsum was a common preservative used in the Roman Empire that acted as a drying agent by absorbing liquids leaving the corpse.[50] Furthermore, molasses also had long been used as a preservative of both food and corpses.[51] As a sugar byproduct, it reduced the amount of bacteria that grew on the body's surface, absorbed moisture, and left behind dehydrated tissues with a texture similar to that of beef jerky.[52] However, for families looking to make more of a statement about their social status and offer additional protection for the corpse, cement ushered in a new era in funerary technology, and the burial vault became the more secure method of body disposition for the well-to-do.

Burial Vaults

While cement was becoming a popular material for coffins, it also began appearing in the construction of burial vaults, a body disposition option for families looking to offer the most protec-

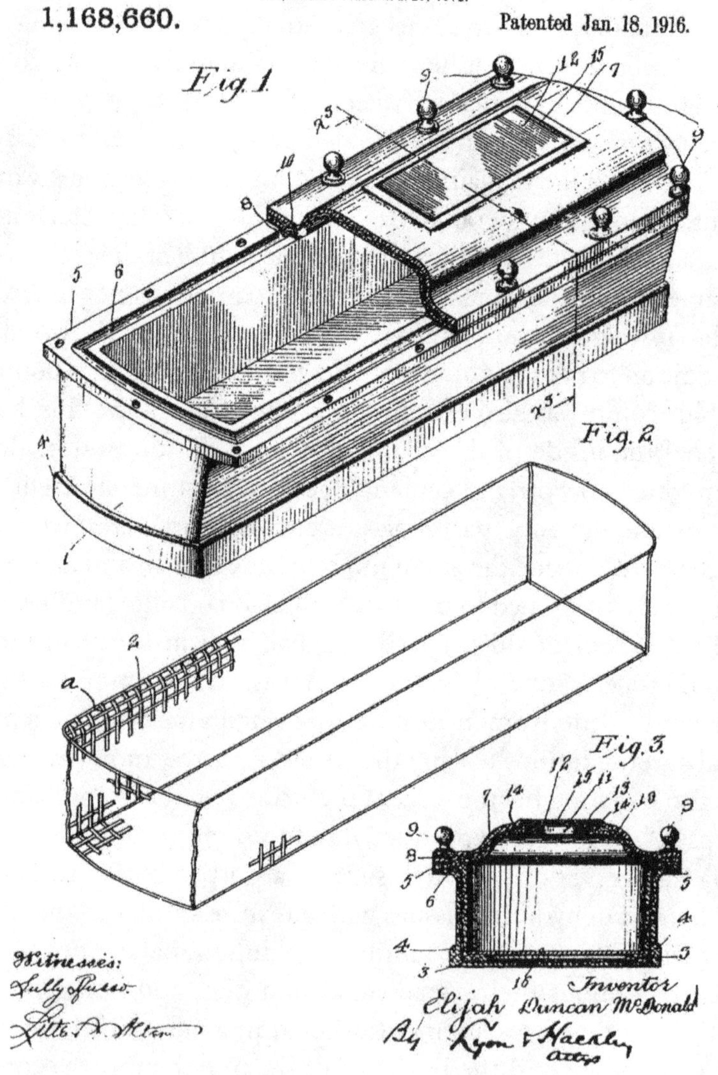

Fig. 10. Elijah D. McDonald, Mexico patent 12,116 (U.S. patent 1,168,660), "Mortuary Box." The hygienic lining described by McDonald would cover the inside of the wire fabric, labeled figure 2 in the illustration. Archivo General de la Nación, Grupo Documental-Patentes y Marcas, legajo 307, expediente 83, September 5, 1911.

tion for the deceased and any valuables buried with the corpse. As one French sociologist has noted, the vault became widespread in the twentieth century with the commercialization of death, since it was no longer sufficient to protect just the corpse; families wanted "to protect what protects the corpse."[53] However, this was not the first time inventors had tried to offer dual-layer protection for corpses. During the nineteenth century in the United States, for example, inventors had tried to line graves and coffins with loose rock, finished stone, or brick.[54] But the common flaw these materials had was that none was hermetic; each contained small spaces through which air, bacteria, and insects could travel. However, by the twentieth century these ingredients had disappeared, as cement was becoming the standard material used by inventors in the funeral industry.

Unlike in the United States or western Europe, burial vaults in Mexico had less to do with repelling body snatchers and more to do with protecting valuables and reducing the threat dead bodies posed to public health. Between 1912 and 1913, the Mexican government granted three patents to Francisco Kassian, a subject of Austria-Hungary who resided in Mexico City. His first patent (number 12,979) was for a burial vault that consisted of a sheet of metal or cement covering an iron frame, for which he received provisional patent status.[55] Traditional burial vaults had coverings made of brick or stone, but as Kassian put it, his vault coverings were more customizable and offered far greater "aesthetic appeal and cleanliness," both appealing characteristics for state officials and elite families (see fig. 11).[56] The sheet of metal or cement placed above the coffin also created a smooth, flat surface that would allow families to stack multiple coffins, a burial option preferred over the invasive and costly procedure of dismantling stone- or brick-covered vaults (see fig. 12). This option would allow living members of the upper classes to claim access to a unique burial method that only they could afford and that would keep members of the same family together for eternity.

Kassian recognized that both Porfirian officials and elite families wanted to customize the deceased's surroundings to reinforce the "natural" order of society. As a result, he submitted another patent that combined his vault coverings with a system of individualized underground burial compartments, each separated by artificial stones impermeable to humidity.[57] He continued to work on improving the construction process of his burial vault, and on July 24, 1913, he received provisional status for burial compartments that combined elements of both his previous patents. This time, he made the compartments from blocks of cement rather than artificial stone and allowed families to fill any empty space with additional cement, which would encase the coffin in a giant cement block (see fig. 13). The result, Kassian explained, prevented unwanted entry and kept humidity and moisture from penetrating the coffin. As he put it, this would "pay tribute to the beauty of returning to dust" rather than allowing the corpse to "marinat[e] in putrefaction."[58]

Yet the use of burial vaults to slow decomposition and protect the dead was not a new feature exclusive to modern societies. Archeologists have uncovered crude burial vaults called barrows that used stones to form protective enclosures for corpses dating back to the third millennium BCE in Scotland and England.[59] While burial vaults have a long history, the unique features of Kassian's vaults were the materials used and the rationale behind their construction. By the twentieth century, cement vaults not only offered families and state officials a way to preserve corpses, but more importantly they safeguarded the health of city residents, all without having to alter the chemical state of the corpse.

Topical Embalming

While burial vaults presented families and state officials with an opportunity to preserve the dead without chemical additives, the obsession with creating a sanitary environment and receptacles for the dead led to the introduction of various pastes or solutions that were placed on the skin of the corpse to ensure long-term protection. Each inventor guaranteed that this topical embalming would protect corpses for a longer period than any coffin or burial vault alone. Topical embalming was also an

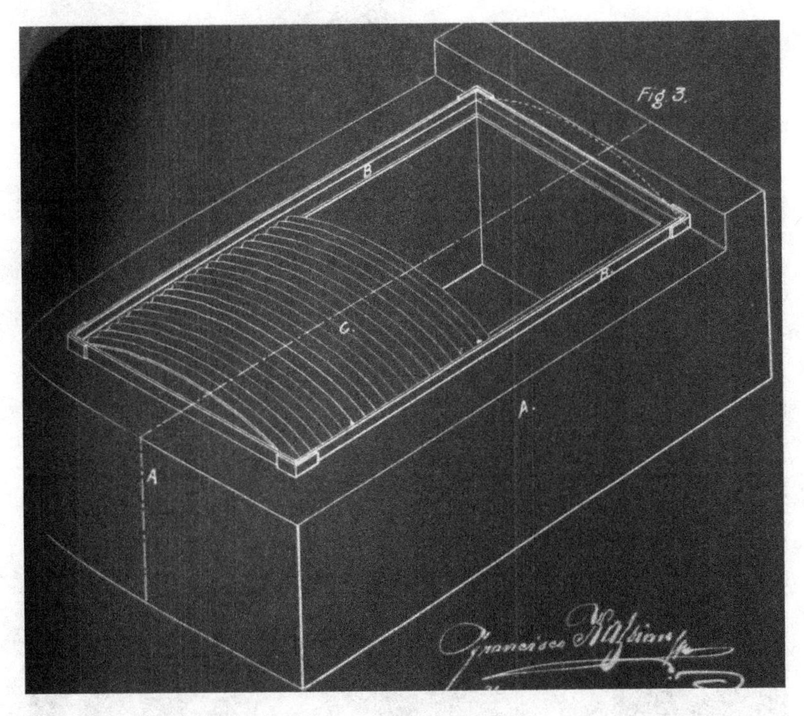

Fig. 11. Francisco Kassian, patent 12,979, "A Covering of Sheet Metal or Rebar for Racks of Tombs." Archivo General de la Nación, Grupo Documental-Patentes y Marcas, legajo 207, expediente 95, May 6, 1912.

attractive option for families because it delivered the same results as chemical embalming but allowed the family to take solace in the fact that medical professionals and undertakers would not poke nor prod the body with needles.

In the United States during the nineteenth century, undertakers used an apparatus known as a corpse cooler to provide topical embalming to families. The cooler was a portable box made from wood or zinc that could be placed on the corpse's torso, head, arms, and legs to slow the decomposition process.[60] Despite its portability and convenience, this method of body preservation had two major problems: the iceboxes were "unsightly and unwieldy," and more importantly, if the undertaker did not drain the melted ice fast enough, the body would absorb the water, thus accelerating decomposition.[61] While it remains unclear whether

FIG. 12. Francisco Kassian, patent 13,134, "Vaults for Tombs." Vaults could hold four coffins, each in its own separate compartment (as pictured in the bottom righthand corner of the design). Archivo General de la Nación, Grupo Documental-Patentes y Marcas, legajo 207, expediente 96, June 19, 1912.

or not the Mexican funeral industry ever used these coolers—it is reasonable to assume that they knew of them, especially in towns along the Texas-Mexico border—newer methods for topical preservation not involving ice began to appear in Mexico during the early twentieth century.

For example, the "Sanitary Urn" was the first topical embalming patent to receive definitiva status on October 1, 1903. Submitted by an inventor who only provided the last name of Jimenez de la Cuesta, the patent's purpose was twofold: to offer a device that

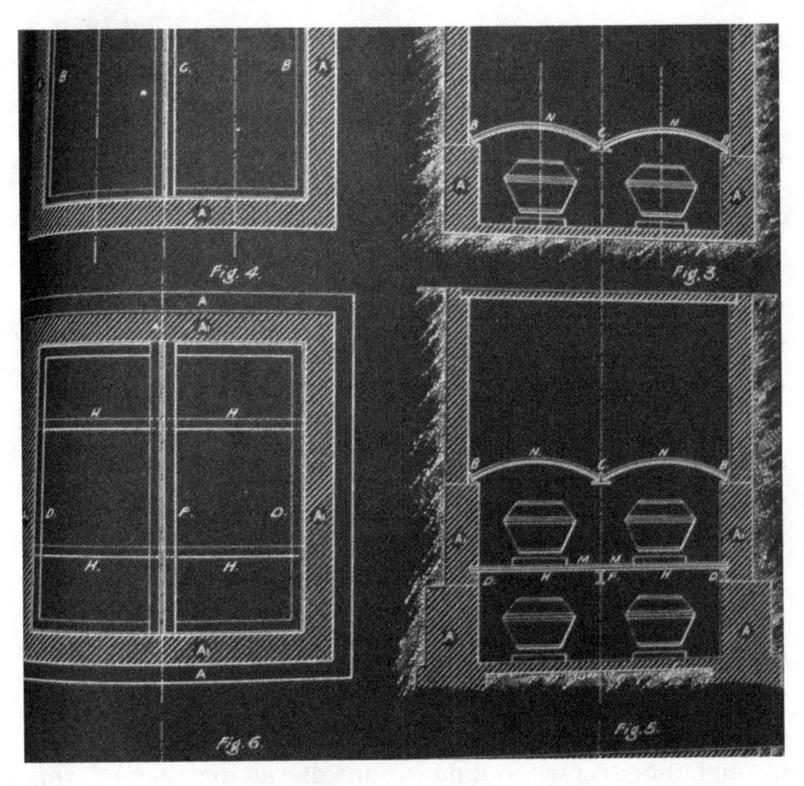

FIG. 13. Francisco Kassian, patent 14,394, "A System for the Construction of Tombs Impermeable to Humidity." This system illustrates how Kassian planned to create an impermeable environment for the corpse. Cement reinforced the individual compartments pictured at the bottom of the design. Archivo General de la Nación, Grupo Documental-Patentes y Marcas, legajo 207, expediente 99, July 24, 1913.

would 1) preserve human bodies and 2) allow for the transportation of corpses in a way that eliminated the potential threat posed to public health.[62] The urn, as Jimenez de la Cuesta wrote, was actually a small generator placed below a soldered box, slightly smaller than a coffin, containing the corpse. "Any disinfecting fluid," he wrote, would produce a cloud of gas that engulfed the corpse, allowing any exposed skin or clothing to absorb the gas. Jimenez de la Cuesta claimed that his invention could preserve the corpse by the time it arrived at the cemetery for burial, without the need to cut the body open. Equally important, his procedure also meant that there was no need for a coffin, which allowed cemetery administrators to bury more corpses in a single plot than the traditional method.

Jimenez de la Cuesta explained the convenience and efficiency that his sanitary urn offered state officials. While the unsoldered side served as a hinged door for placing the corpse inside, liquids could not enter or escape. The soldered box also contained a baseboard made of iron, with a waffle-shaped screen that allowed the gas to pass through to disinfect the corpse. For state officials, the device offered two intriguing features that supported their desire to make the city appear modern through the elimination of the threat decomposing corpses posed to public health. First, the device came with its own box for the corpse. This gave it a more utilitarian feel, which made it more attractive to state officials disposing of myriad lower-class bodies, since it saved them from having to buy hundreds of coffins. Second, the device's portability would allow officials to use it with any vehicle; as Jimenez de la Cuesta promised, "even the most infected body could pass through the street without posing any danger to passersby." This method of bodily disposition provided officials with the opportunity to sanitize corpses in a way that would not threaten the health of citizens or require additional space in cemeteries since burial with a coffin would be unnecessary. Cemetery workers could now bury these preserved corpses without coffins and be confident that public health was safe. As Jimenez de la Cuesta put it, the urn was a "huge benefit to humanity and powerful facilitator of health" for state officials and city residents worried about the potentially devastating effects that decomposing bodies could have on public health.[63]

The "Sanitary Urn" was just the beginning of a flurry of funerary patents that promoted topical embalming as the solution for protecting the deceased and creating a healthier city environment. American physician Marshall Devereaux Johnston, hailing from Bisbee, Arizona Territory, received definitiva status on February 17, 1906, for a patent called "A Procedure for Conserv-

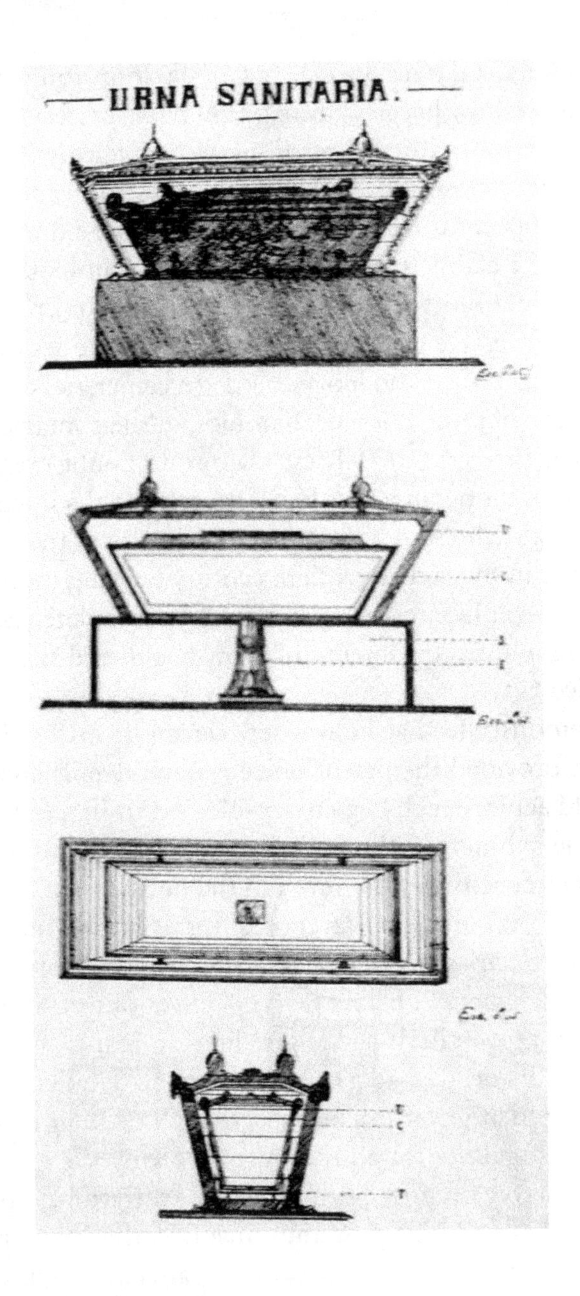

FIG. 14. Jiménez de la Cuesta, patent 3,443, "Sanitary Urn." The small generator below the coffin produced a cloud of preservative gas that surrounded the body. Archivo General de la Nación, Grupo Documental-Patentes y Marcas, legajo 207, expediente 76, October 1, 1903.

ing Cadavers." Like Jimenez de la Cuesta, Johnston made bold claims about how his procedure would offer "perfect health" for corpses. The procedure delivered "perfectly hygienic results" and permanently conserved the body's features and expressions so it would appear "as natural in death as it appeared in life." Furthermore, he claimed his procedure would eliminate the need for traditional burials, as the decomposition of corpses was no longer harmful. He believed that burial in cemeteries was a relic of the past, as corpses no longer needed a cemetery. For him, the elimination of cemeteries meant avoiding "the contagion to the water that they caused."[64] These claims, no doubt exaggerated, were important because they likely appealed to the intentions of state officials who sought to protect the health of the city. With cemeteries in the early twentieth century burying more people daily than ever before, the opportunity to completely abandon cemeteries piqued the interest of many health and state officials in Mexico City.

To demonstrate that he was serious about his bold claims, Johnson provided the patent office explicit details about how he would achieve such hygienic results. According to his application, he submerged the body in a saltwater solution, which allowed electricity to flow through the body easier. Afterward, he covered the corpse in a layer of "silver or carbon nitrate" that acted as an electrical conductor and natural antimicrobial agent.[65]

While it remains unclear how effective Johnston's method was, the approach itself had antecedents in France, where electroplating of corpses had gained a following decades earlier. In 1857 the French government had granted a patent to Eugene Theodor Noualhier for a procedure involving the application of a large layer of "silver nitrate that penetrates the pores of the corpse's skin." This procedure involved the submersion of the body in a bath of copper sulphate, afterward applying silver nitrate to the skin, which along with the galvanic current "established a metallic deposit of copper of the requisite thickness, the result being a metallic mummy."[66] A similar procedure had appeared in the United States during the nineteenth century that prevented decomposition by using various solutions to soak the body, "rendering its contents impervious to all forces of decomposition."[67] The popularity of electricity as a preservation method remained ever-present throughout the nineteenth century and into the twentieth century. In 1935 American Levon G. Kassabian would receive patent rights for his unique "Method of Preserving Dead Bodies." Unlike Noualhier's procedure, Kassabian's required injecting the corpse—before submerging it—with a solution of zinc chloride, sodium salicylate, phenal crystals, and water. After several injections of this solution, Kassabian pointed out, the body's ability to act as an electrical conductor increased greatly. He would then place a thin layer of paraffin wax and copper wire over the body, which he believed would allow electrical current to move uninterrupted throughout the corpse and thus preserve the deceased.[68]

While electroplating was—and would continue to be—a popular method for corpse preservation, Mexican medical surgeon Agustín Domínguez Fagle offered state officials another alternative method of body preservation. While it achieved the same results as electroplating, it did not require soaking or brushing the corpse with silver nitrate. On May 3, 1907, Domínguez Fagle received provisional patent status for Aegecil, a topical embalming ingredient placed inside coffin cushions that the skin of the corpse would absorb. He made Aegecil from ingredients commonly found in medical labs—such as thymol, nitric acid, lime hydrochlorate, and aluminum sulfate—as well as naturally occurring substances such as oak charcoal, sawdust, balsam of Peru, French fine grass, powdered tar coal, and eucalyptus wood chips or leaves.[69] Mixed together, Domínguez Fagle claimed, these ingredients had dehydrating properties that would remove any remaining liquids from the corpse.

He also provided the patent office with the specific measurements to create Aegecil. Two formulas were involved: one for the cushion filling and one for the cushion covers. The cushion filling was a combination of four kilograms of pine sawdust, four kilograms of eucalyptus leaves, four kilograms of French fine grass, four kilograms of oak charcoal dust, two kilograms of coal tar dust, and six kilograms of lime hydrochlorate (calcium chloride). Domínguez Fagle explained that this mixture would protect the body to such a high degree that "one will not lose any part of the human skeleton, unlike other procedures." The second formula called for a solution made from two liters of 1 percent thymol (an antimicrobial solution), one kilogram of nitric acid, one kilogram of aluminum sulfate, one liter of formol (an early version of formaldehyde), and five hundred grams of black balsam of Peru. Afterward, Domínguez Fagle soaked the cushion covers in the solution for five to ten minutes, then air-dried them. Once dry, they would "indefinitely conserve the mortal remains and mummify them." According to Domínguez Fagle, this procedure offered more protection than other forms of topical preservation and made the process far more efficient, since no injections of the corpse were required.[70] Whether these cushions actually slowed the body's decomposition and protected the health of living residents remains unknown.

Nonetheless, by summer 1907 Domínguez Fagle's procedure appears to have set off a firestorm of professional competition among inventors of funerary technologies over how best to preserve the corpse and protect public health. Three months after Domínguez Fagle received his patent, another inventor received definitiva status for a patent that would embalm all organic bodies—both human and animal—without removing any organs. Giovanni Chiarelli Fu Giovanni, the man behind the procedure, hailed from Genoa, Italy. His embalming procedure required the corpse to soak in one of three chemical formulas for four or five days depending on, as he put it, "the quantity of flesh."[71] The first solution contained a mixture of six grams of an unnamed corrosive sublimate mixed with one liter of alcohol, the second used 100 percent pure Lysoform (made from liquid soap and formaldehyde), and the third used a combination of equal parts Lysoform with the first solution of corrosive sublimate. After soaking, he placed the corpse on a bed, table, or coffin for several hours "to dry slowly without giving off any fetid odors." Regardless of which solution mixture he used, the results were the same: one could manipulate the corpse into any desired position and it would remain "perfectly rigid" forever. According to Fu Giovanni, the procedure was so effective that a simple observation would indicate whether the corpse posed a threat to the living world: if "no flies appeared on the body," there was no threat.[72]

Fu Giovanni's patent was not the only one to present a method for the preservation of corpses by soaking them in a solution. Antonio Subira, a medical doctor from Barcelona, Spain, received provisional status for a procedure that consisted of "showering, painting, or soaking" the corpse in an unspecified substance that had dehydrating properties. Subira promised that this would remove all water from the body and thus delay decomposition, which reduced the potential threat posed by the corpse. While Subira did not provide the specific formula he used for the solution, he did explain that he used chemicals such as glycerin, calcium chloride, caustic soda, methyl alcohol, and potassium nitrate in "various proportions" to complete his procedure.[73]

Despite supposedly being modern, the actual process of topical embalming was not as new as state officials or inventors wanted to believe. The approach dates back millennia to the Egyptians, who eviscerated the corpse before embalming the internal organs and wrapped the corpse in a shroud soaked in oil and spices to reduce decomposition and protect public health.[74] Ancient Ethiopians covered corpses in plaster, while Persians, Syrians, and Babylonians submerged corpses in honey to preserve them. However, sometimes these processes also involved disemboweling the corpse and filling its cavities with substances such as frankincense or aniseed to reduce the stench of decomposition. Even in sixteenth-century England, barber surgeons embalmed the corpses of the royal family by removing their organs and filling their cavities with cotton soaked in a mixture of aloe, myrrh, rosemary, and other spices.[75] Nevertheless, the introduction of chemical substances and use of new technologies separated the new from the old and made these patents appealing to state officials looking for potential solutions to their public health problems.

Arterial Embalming

Unlike topical embalming, arterial embalming was far more invasive, and early versions delivered less-than-desirable results. Yet it would become a popular preservation method in early twentieth-century Mexico City for physicians and funerary professionals who wanted a fast and efficient way to protect the living from potential diseases found inside dead bodies. Injecting organic material with preservative solutions was first attempted in 1663, when Robert Boyle, an experimental philosopher and arguably the most influential figure in the emerging scientific culture of late seventeenth-century England, published the results of experiments he had conducted on animal preservation through vascular injections made from wine and other spirits.[76]

Yet the transition from animals to humans did not occur until the eighteenth century. Then, Scottish anatomist and physician brothers John and William Hunter injected various chemical solutions into the arteries of corpses to determine their rate of decomposition. Meanwhile, in the United States, an embalming procedure through arterial injection did not appear until 1856, when J. A. Gaussardia of Washington DC received a patent for washing the corpse in chemicals, covering it in oils, and injecting it with an arsenic-alcohol mixture before electrification, all of which would preserve the corpse.[77] But the process of arterial embalming remained a distant secondary option for corpse preservation.

However, the U.S. Civil War ushered in a new era for arterial embalming experiments in order to keep up with the staggering number of corpses collected daily. The person responsible for popularizing modern arterial embalming during this time was "Doctor" Thomas H. Holmes. Holmes, who advertised himself as a doctor of corpse preservation, would eventually claim to have embalmed more than four thousand cadavers in four years. Despite his title, Holmes never graduated from New York University medical school, although he did attend briefly in the 1850s. Although Holmes never revealed what his exact embalming formula contained, experts believed he used a combination of zinc chloride and arsenic, a popular form of body preservation at the time.[78] Without having graduated, it was highly probable that he did not know how to perform actual arterial embalming. Rather, he could perform a more crude and thus easier embalming method known as cavity embalming that relied on inserting trocar-cannulas filled with preservatives into the major cavities of the body. He flaunted his medical and embalming expertise, which appealed to families of fallen officers who had the financial resources to pay $100 ($1,580) for his services. His was a unique service at the time. Many of the soldiers were far from home and often were never to return alive, and the process allowed for preservation and shipment that enabled proper burial. During the late nineteenth century, most embalmers only swabbed the corpse with embalming fluid since injecting the corpse meant using tools that were costly and not readily available in all markets. Moreover, few embalming schools existed. For example, in the United States, there were only two schools of embalming—the Cincinnati School of Embalming and the Rochester School of Embalming, both founded in 1882.[79] In reality, many town undertakers throughout the United States had very limited anatomical knowledge, which made arterial embalming a less popular option since it required a more sophisticated understanding of the human body.

Despite the rise in the number of embalming schools in the United States by the turn of the twentieth century, the curricula of these institutions continued to focus almost exclusively on cavity embalming rather than arterial. While arterial embalming preserved bodies longer, it also required an advanced understanding of human anatomy, especially of the vascular and arterial systems, which most embalmers did not possess. The only requirements for "belly puncturers," as they were popularly known, was to cut open the body, remove the organs, sew the incision shut, and dress the body to hide poor stitching. Furthermore, cavity embalming was a less effective method of preservation because it only protected areas that received the most preservative fluid.

Its ease, however, saw it become popular among arterial embalmers, who began to include cavity embalming as part of their procedure to ensure total preservation of the body.[80]

Nevertheless, by the early twentieth century, arterial embalming remained a new process in Mexico. Patent records indicate that the first time arterial embalming gained attention was in 1908, when Mexican inventors Carlos C. McRae and Antonio J. Ogazón received provisional status for their patent "An Apparatus for Injecting Cadavers." McRae and Ogazón emphasized that their device was both "new and useful" for disposing of corpses. Their embalming process, the duo explained, started with the embalmer inserting several trocars, medical instruments with sharply pointed ends used inside cannulas (hollow cylinders), into blood vessels or body cavities to extract blood, fecal matter, and gases from the body (see fig. 16). Once this matter was extracted, the embalmer inserted several needles into the corpse to fill it with preservatives and antiseptic liquids. All of this sanitized the body and reduced the potential threat that decomposition posed to the environment. While "Doctor" Holmes could embalm three corpses per day, McRae and Ogazón's machine could embalm ten to twelve corpses per hour.[81] According to McRae and Ogazón, their embalming machine would need less than five minutes to embalm one corpse. The efficiency demonstrated by this technology was an important tenet of Porfirian modernity, supported by the construction of the machine itself, which included two different-sized chains to move the gears and power the needles and tubes in a way that would emphasize "strength and perfect uniformity."[82] The machine would serve as a symbol of strength for both public health and technology. If it were adopted, the threats posed to public health would diminish and showcase how science and technology were triumphing over death in Mexico.

The embalming machine submitted by McRae and Ogazón, however, was not the only one of its kind to capture the attention of Porfirian state officials. Cuban physician José María Addis had the opportunity to share his embalming machine before an

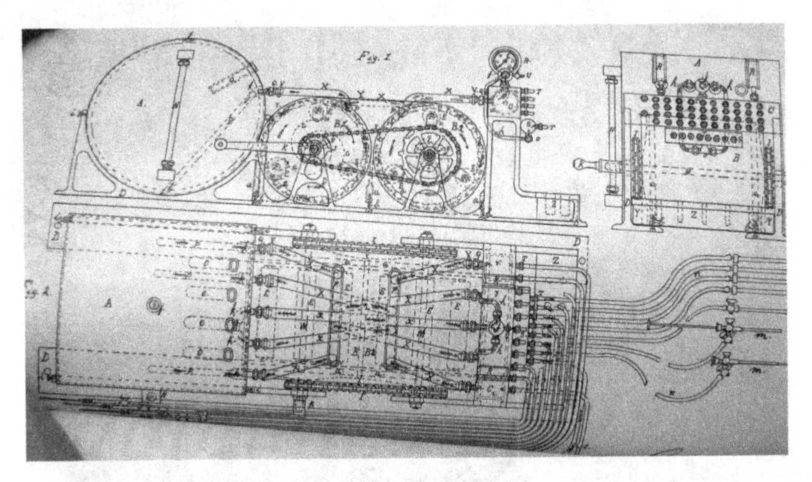

Fig. 15. Carlos W. McRae and Antonio J. Ogazón, patent 8,047, "An Apparatus for Injecting Cadavers." The two gears powered the seven needles at the bottom right of the drawing, which allowed embalmers to extract fluids from the corpse and inject it with preservative liquids. Archivo General de la Nación, Grupo Documental-Patentes y Marcas, legajo 207, expediente 85, May 29, 1908.

audience of state and medical officials at the National School of Medicine in Mexico City on March 8, 1909.[83] The event, covered in the March 9 issue of *El Heraldo* newspaper—a large Catholic opposition paper not favored by the state—shed light on the Díaz regime's struggles with the disposal of corpses and methods for protecting and improving public health in the city.[84] His embalming procedure, which included a detailed pamphlet in English and Spanish, is worth exploring in more detail. In particular, it offers a glimpse into how embalming captured the attention of Mexico City's residents, providing readers of *El Heraldo* disturbing images of how modernization could improve public health in the capital and how modern science and medicine could forever change the burial process.

The pamphlet Dr. Addis presumably handed to audience members at the demonstration began with a short history of arterial embalming, making sure to point out the important legacy left

UEDE SER EMBALSAMADO UN CADAVER EN DOS O TRES MINUTOS SOLAMENTE

NOTABLES PRUEBAS EN LA ESCUELA DE MEDICINA

tuvieron verificativo en el
de la Escuela Nacional
las pruebas practi-
embalsamiento rápido de
por medio de un proce-
inventado por el sabio
don José M. Addis.
gran espectación por presen-
pruebas, en vista de que
procedimientos que se usan
para la conservación de
requieren varias horas de
y el empleado por el Doctor
verifica en el brevísimo es-
3 á 4 minutos.
once de la mañana, la sala
se encontraba muy con-
por doctores, estudiantes y
así como por represen-
varios periódicos de la

la llegada del señor
del establecimiento, pero de-
indisposición no pudo asis-
en su lugar al Doctor
Secretario de la Es-

principiar las experiencias,
presentó su Excelencia el
de Ranuzzi Segni, Mi-

FIG. 16. "How One Can Embalm a Cadaver in Only Two or Three Minutes: Notable Experiments at the School of Medicine," *El Heraldo*, March 9, 1909. In the photograph, the hand of Dr. Addis (the man in the white lab coat) is on top of his embalming machine, which is performing a muscular embalmment through the major muscle groups of the cadaver lying on the table. Archivo Histórico de la Ciudad de México, Fondo-Ayuntamiento de México/Gobierno del Distrito Federal, Serie-Panteones, caja 3475, expediente 28, July 7, 1918.

by famed eighteenth-century Scottish anatomist and physician William Hunter. The narrative quickly turned to promoting the importance of total embalming (arterial, cavity, and muscular) as a process that "protected the living from possible infection." He assured readers that this was the only method that would remove germs from corpses, which frequently leaked into the soil and carried "contagion to the living, through water and sometimes through food." The result of his method, Addis proclaimed, was a preserved corpse that no longer posed a threat to the living world, which was extremely appealing to state officials and the medical community looking to improve public health and to well-to-do families seeking eternal protection from insects and decomposition.

According to Addis, achieving these results was not difficult; operating the machine was quite simple. As he put it, even the most "inexperienced operator will easily find the left ventricle of the heart," the first step in his embalming process. While Addis preferred arterial embalming to cavity, he realized that "if a perfect embalming is desired," he would have to include cavity and vascular injections also. Using air pressure from his machine, Addis could complete total embalmment in five to ten minutes, spreading five quarts of embalming fluid into the cavities and muscles.[85] Thus, his machine completed the embalming process in such a short time, delivering unmatched results that safeguarded public health in Porfirian Mexico City.

Although Addis had extolled the ease of using his machine, reality was far different. Operating the embalming equipment required a far more delicate touch than he posited, since there were various pressure settings that accelerated or decelerated the process. Nevertheless, Addis suggested that inexperienced machine operators practice by filling the preservative tanks with water instead of embalming fluid and injecting "a dog or other big animal" before moving on to a human. Practicing individual technique and adjusting the pressure settings would ensure that they were far less likely to "seriously bloat the corpse or even rupture its skin."[86] Such an error could tarnish one's social and professional reputation, opening both funerary professionals and state officials who licensed them to possible litigation. Moreover, public mistrust could deliver devastating consequences in the state's desire to capitalize on the population management techniques prevalent during the Díaz presidency.

After becoming familiar with the settings, the next step was muscular embalming. The first phase of the process began with the eyes. The embalmer inserted one small needle into the corner of each of the corpse's eyes, and when the embalmer reached a depth of 6 inches, he turned on the machine to deposit the required 16 ounces of fluid in each eye. Next, he inserted one needle into each lung, depositing 11 ounces of fluid in each. Afterward, the embalmer placed 3 needles into the abdominal cavity and injected 23 ounces of fluid, then inserted a trocar into the left ventricle of the heart and pumped 60 ounces of fluid into it.[87] Once these steps were completed, the embalmer could be relieved that he had eradicated any disease found in these parts, according to Addis.

The second phase of muscular embalming involved inserting two needles on each side of the pectoral muscles, one needle in each shoulder, and two needles in the throat—the area that decomposed the quickest—each of these areas receiving fourteen ounces of embalming fluid. The third phase involved inserting four needles into each arm and injecting twenty-two ounces of fluid. The fourth phase placed one needle per buttock and one into the "fleshy part of each thigh," injecting a total of forty-seven ounces of fluid. The last step of the embalmment process placed three needles in each leg and one in each foot for an additional twenty-two ounces of embalming fluid.[88] Now, the embalmer would have a partially preserved body but one that still required more work.

The next step in the process was cavity embalming. Addis described this as using "three quarts of fluid in the thoracic and abdominal cavity." Once the embalmer completed this process, Addis explained, the corpse could return to a "normal state," even if he or she were a victim of contagious disease, drowning, or a train accident.[89] This dual process of embalming was important for eradicating disease from the body and protecting public health in the city.

There was one disease thought that undertakers and medical officials continued to struggle with: tuberculosis. Addis believed, however, that his method could solve this struggle. According to him, normal embalming in tuberculosis cases required the embalmer to inject the carotid artery and bronchial tubes with fluid. But this procedure caused more problems than it solved, since inserting fluid into these areas was a delicate process, and thus internal leakage was quite common. To find the origin of the leak, the embalmer would repeatedly cut open the corpse, starting with an incision in the carotid artery, both femoral arteries, the windpipe, the eyes, and finally, the usual location of the tear, the abdomen. These actions, Addis exclaimed were "profane and extremely repugnant," leaving behind a semimutilated corpse.[90] His machine could not only protect against tuberculosis but, as he assured state officials, it would neither cause leaks nor leave behind carved corpses which had served as unflattering symbols of the lackluster scientific modernization process in Mexico.

In particular, Addis explained that his machine and embalming process could return any corpse to a more natural appearance. His technique—injecting the eyes first—eliminated a common problem associated with embalming. Rather than the eyes "swelling like hen's eggs," Addis's process gave them a more lifelike appearance. Additionally, his process allowed him "to clear the blood from the head and rid it of its discoloration," another characteristic associated with typical embalming procedures, which often left corpses looking unnatural and awkward. He also wrote that by injecting the left ventricle, "preservative fluid can circulate through all of the arterial system," which would provide a more lifelike appearance for the rest of the corpse. Addis concluded his pamphlet by reasserting the fact that "even an illiterate person" could operate his machine. This statement—while an exaggeration—was appealing to state officials, who would frequently hire such people to work with or near bodies in positions such as gravedigger or morgue assistant.[91] But more importantly Addis's machine proposed to offer state and medical officials an opportunity to solve the city's corpse problem using machinery and man, an approach befitting a modern, urban capital.

Nonetheless, it appears that state officials did not adopt Addis's embalming method since he does not reappear in the historical record until November 6, 1918. That month, Addis wrote a letter to the municipal president of the Federal District, José María de la Garza, to offer his embalming expertise for the recent victims of the Spanish influenza that had swept through Mexico.[92]

This global pandemic hit Mexico hard. Influenza was a familiar disease that left its victims bed-ridden for several days or weeks before their health returned. Spanish influenza, however, was an unknown strain that carried pneumonic complications.[93] While the official statistics of Spanish influenza–affected residents of Mexico City are controversial—some estimates place the mortality rate at 2 percent, while others have it closer to 0.7 percent—the impact the disease had on medical and state officials is not. With a population of between 720,000 and 800,000, influenza resulted in anywhere from 5,000 to 15,000 deaths—and corpses—in just a few months.[94] Such large numbers forced state officials to store the dead bodies in monstrous piles at the former Belén Prison, and Addis believed his method would help "sterilize" the situation.[95] Such a sight surely upset revolutionary state officials, who had continued to define modernity in Mexico City along Porfirian ideals, especially those related to protecting public health.

Nevertheless, before Addis would commit to helping eliminate the risks to public health posed by the decomposing bodies, he wrote a series of contractual demands he expected the government to meet. First, workers had to take the corpses out of the large piles at Belén and divide them into smaller piles and separate them into even smaller piles according to body size. Second, the government had to pay for the antiseptic fluid that his machine would need to embalm the hundreds of corpses. Third, the government would pay him a fee of 1.50 pesos (most likely between $45 and $74.10) per embalmed corpse, which he acknowledged was a bit pricey—but would allow him to pay his medical assistants.[96] Fourth, he would need the corpses dressed before he would begin the arterial embalming process. Moreover, if the corpses had poor vascular systems, meaning they were too far along in the putrefaction process to preserve, they would only receive cavity and muscular embalming. Finally, Addis promised the municipal president that his method was the best option, since he would be able to embalm "no less than 200 cadavers per day" for the length of his three-month contract, an amount that tra-

Fig. 17. Dr. J. M. Addis, "Cranial Cavity Injection." Dr. Addis is injecting preservative fluids into a severed head of an unknown man. The various bottles and tubes in the foreground delivered preservative fluids to the cranial cavity, which included a nasal cavity insertion. Archivo Histórico de la Ciudad de México, Fondo-Ayuntamiento de México/Gobierno del Distrito Federal, Serie-Panteones, caja 3475, expediente 28, July 7, 1918.

ditional burial methods could never match.[97] Addis hoped that this time state officials would see the benefits his machine could bring them in their quest to protect the city from the threat of decomposing corpses.

Thirteen days later, Municipal President José María de la Garza responded to the request. He informed Addis that he had received his proposal and had hired his own medical expert to review the legitimacy of such bold claims. Addis did not impress the municipal president's expert, R. Riverell, who concluded his review by stating that while the embalming process was "nice, it did not

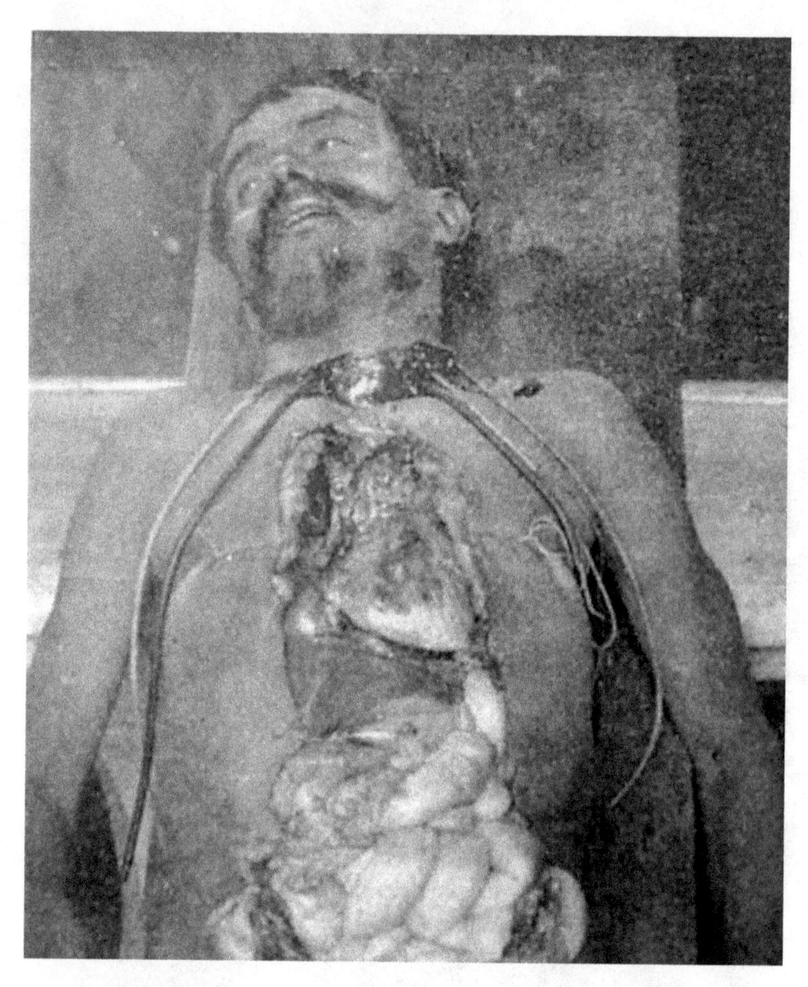

FIG. 18. Dr. J. M. Addis, "Left Ventricle Injection." Dr. Addis has cut open the cadaver to illustrate how the preservative fluids move into the left ventricle. The tubes in the picture connect the arteries in the neck with those in the chest. Archivo Histórico de la Ciudad de México, Fondo-Ayuntamiento de México/Gobierno del Distrito Federal, Serie-Panteones, caja 3475, expediente 28, July 7, 1918.

have any influence against disease." However, despite the review, de la Garza informed Addis that he would grant him a contract if he would agree to pay for his own antiseptic or use "a cheaper antiseptic" because he believed that there was no difference in

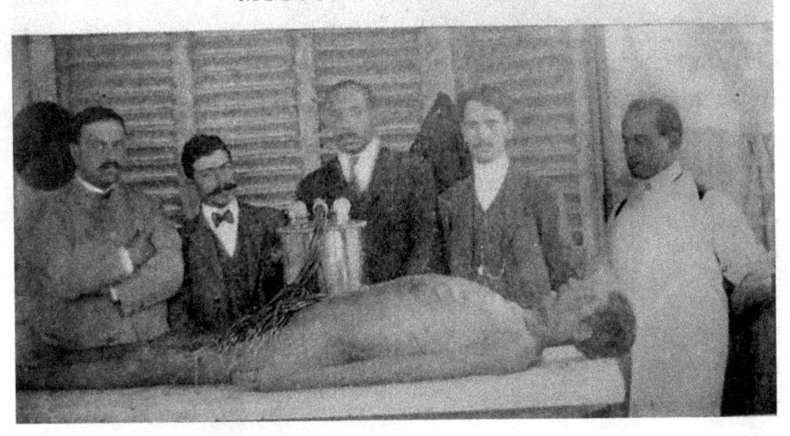

Fig. 19. Doctor J. M. Addis, "Muscular Embalming." This picture shows how Dr. Addis performed muscular embalming. According to the pamphlet, the tubes connect preservative fluid to the major muscle groups. Archivo Histórico de la Ciudad de México, Fondo-Ayuntamiento de México/Gobierno del Distrito Federal, Serie-Panteones, caja 3475, expediente 28, July 7, 1918.

the protective abilities between the two versions except their price.[98] Alas, surviving records do not reveal whether or not Addis accepted the condition that he would have to either pay for the antiseptic or use one below his medical standards. Regardless, concerns about the standard of public health remained integral features of the Mexican government, even after Porfirio Díaz had fled the country in the early years of the Revolution.

Crematory Ovens

Officials' need to solve the issues surrounding the public health conditions in the city caused them to explore the potential that fire could offer a cleaner disposition method for corpses, especially those of the poor. While embalming had gained popularity among state officials in the early twentieth century, cremation captured the imagination of state officials and supporters in Mexico, who for years had looked for ways to dispose of the dead in an efficient and sanitary manner. Cremation presented officials with an opportunity to avoid machines, antiseptics, and even cemetery space. The introduction of the Marburg Crematory Oven in October 1907 offered a more efficient and modern method for reducing Mexico City's escalating corpse problem. Additionally, the crematory oven also provided city cemetery administrators at places like Panteón Dolores an alternative method for disposing of the corpses of the urban poor, since cemetery space was at a premium.[99] The oven helped Porfirian state officials reinforce class divisions: as a disposition method, it would only be used for the bodies of those considered unwanted. In the end, such a "safe" and "healthy" option would allow the city to reach a level of modernity never seen before.

Negotiations between Guillermo Beltran y Puga (the director of public works in the Federal District) and representatives from the Caesar J. Marburg Company concluded on October 12, 1907. The Marburg Company would build two crematory ovens in Panteón Dolores, one that could incinerate one corpse every two hours and another that could fit three to five corpses at once and incinerate them in three to four hours. The details of the contract also specified that the Marburg Company would supply "all material and iron" needed for the construction of the ovens. Meanwhile, the Mexican government would pay for "all costs of preliminary construction" as well as the transportation costs of moving the materials from Veracruz to Mexico City using one of two state-owned railroad companies (Mexican Railroad and Interoceánico). But if Marburg were unable to complete construction by July 1, 1908, then the company would pay Mexico's Treasury Department 30 pesos per day ($908) in fines until its completion. However, according to the contract, if the ovens were finished on time and passed inspection, then the government would pay the Marburg Company a 40,000-peso bonus ($1.21 million) for its contribution to efficiency and promotion of public health in the capital.[100]

The director of public works alerted the Marburg Company there were seven conditions that would allow the government to void this contract without owing the crematory company any money. To satisfy state officials' needs to protect public health, the ovens had to demonstrate the following characteristics. First, the ovens had to cremate corpses in a specified time—one corpse every two to three hours in the smaller oven and three to five corpses every four hours in the larger oven. The ovens also had to produce white ashes and bone fragments measuring "no more than two centimeters in length." Additionally, the amount of fuel each oven used to incinerate a single body could not exceed one hundred kilograms. The gas produced by cremating corpses also had to be "neutral"—a phrase that state officials failed to explain further but that most likely meant not harmful to city residents and cemetery workers who would ingest the smoke on a daily basis.

Conditions five through seven of the contract hinged on materials and transparency in the construction process. If the Marburg Company were unable to obtain the materials needed to complete the construction within three months or if the ovens failed to meet the minimum corpse capacity, the company would receive no compensation. If the director determined that the iron and other materials used during construction were "not of good quality," the company would get nothing. The government's last stipulation explained that if Marburg did not provide public works officials with a written notice of the completion of the ovens, then the company would go home emptyhanded.[101] Such stringent clauses in the contract were the result of the tremendous power the Department of Public Works wielded over people interested in investing in Mexico City's infrastructure, especially related to improvements in public health. As the director of such a department, Guillermo Beltran y Puga could choose anyone he wanted to complete this project. As such, it was an opportunity that stood to make someone a financial and social success in Porfirio Díaz's Mexico for saving public health from further ills.

Not surprisingly, with so much money and prestige on the line, the Marburg Company completed the ovens before the deadline, satisfying all of the requirements imposed by the director of public works. While completing the ovens by July 1, 1908, may have seemed like a normal deadline for a project of this magnitude (eight months), the actual reason why Beltran y Puga pushed so hard to expedite the work was because the city was running out of burial space in Panteón Dolores. Cemetery administrator Alberto Hope had informed the Department of Public Works in early December 1907 that corpses had overrun his cemetery, where almost 22,000 bodies had been buried between January 1906 and December 1907.[102] To create the space required for such large burial numbers, the cemetery administrator had ordered the exhumation of corpses in individual graves with seven-year burial terms. For example, the 1887 rate for a ten-year burial in a third-class grave was 20 pesos ($758); for perpetuity it was 100 pesos ($3,790).[103] Even in 1919, the rate for a seven-year burial in a third-class grave was 15 pesos ($231), while for perpetuity it was 75 pesos ($1,150).[104] As a result, the exhumations only affected individuals from average or below-average socioeconomic backgrounds, who were far more likely to pay for a term burial with lower upfront cost. For the well-to-do, perpetuity was affordable and the only valid option for the deceased.

Hope also pointed out that if the number of burials remained at the current rate, then he and his department would have to consider expanding the cemetery. When the federal government purchased the land that became Panteón Dolores in 1879, it was more than 276 acres (about 1.1 million square meters).[105] According to Hope's calculations, which he based on a seven-year burial term and normal mortality rates, the cemetery would need an additional 170 acres (almost 700,000 square meters) to accommodate the growing number of corpses buried in 1908. Expanding the cemetery would allow for one meter of space between each grave as well as the option to expand the width of cemetery roads to accommodate the carriages transporting corpses and thus present a more attractive cemetery.[106] But if expansion was not an option, Hope had another plan. If state officials reduced standard burial from seven to five years and allowed the cemetery to keep the ground "continuously wet, through rainfall or manual watering," that would accelerate the decomposition of corpses buried with or without coffins, only requiring a cemetery footprint of almost 500,000 square meters (44 percent of the current cemetery's size), slightly smaller than his initial calculations.[107]

Despite the plan's promise, state officials failed to adopt Hope's recommendations—most likely because such large tracts of land were difficult to find in an expanding and overcrowded Mexico City. Additionally, other bodily disposition methods such as cremation were more practical and in their minds also represented the cleanest and most modern disposal method for a capital seeking to improve poor hygienic conditions and improve its image.[108] For elite and middle-class residents who blamed the poor for the city's appalling health, cremation fed the flames of modernity with the bodies of those they considered inferior.

Nevertheless, one of the major reasons why the poor were more hesitant to embrace cremation for their deceased was the important role religion played in the construction of their culture. In particular, elite and middle-class residents in favor of cremation often cited popular Roman Catholic superstition as the reason why the urban poor failed to embrace the warmth of modernity embodied by the crematory ovens.[109] Similar debates had also appeared around the world, for example, in the late nineteenth-century Austrian capital of Vienna. In 1873, in the middle of the World Exhibition in Vienna, a cholera epidemic created a debate between city and church officials about what was the best and most hygienic method of body disposition for the dead. Coincidentally, a German inventor, Werner von Siemens, had provided Viennese officials with an opportunity to use his cremation apparatus. Its appeal to modernizing government officials outweighed any concerns expressed by the Catholic Church, especially since cremation had both "sanitary and utilitarian advantages," saving scarce burial ground in city cemeteries.[110] Liberal newspapers in Vienna, as well as state officials, promoted von Siemens's machine and the process of cremation as an invaluable hygienic tool that symbolized the progress underway in Vienna.

Yet the Catholic Church remained vehemently opposed to the use of cremation in Austria. Despite the establishment of the first European crematorium in Milan, Italy, in 1876—followed by Gotha, Germany, in 1878 and throughout England in 1902—the church and conservative political parties firmly opposed cremation for decades, until church and state officials were able to reach an agreement legalizing cremation in Austria by 1923.[111] The stance against cremation in Mexico also had a long history. The Catholic Church's attempts to ban cremation had taken place as far back as the sixteenth century, when church officials sought to eliminate the cremation practices of Mexican indigenous populations and substitute Roman Catholic burial rights in their place. For example, church officials were especially weary of Aztec practices that used cremation since it was such a foreign concept to church officials at the time. For the Aztecs, however, cremation held significant spiritual value as the only way to facilitate the movement of the heart through the underworld and for the practical purpose of fertilizing crops.[112] While cremation was popular among indigenous groups in Mexico, Spaniards, creoles, and mestizos preferred burials, which became a fashionable way to reinforce social hierarchy from the colonial to the modern era.

During the presidency of Porfirio Díaz, death was an influential tool in the construction of the country's official history.[113] In particular, burials were useful for certain types of people (the well-to-do) and cremation for others (the poor). This segregation even in death helped Díaz extend his power. While Porfirio Díaz and his científicos had tried to eliminate the influence the church held over the poor, it was an impossible task. The several-hundred-year stranglehold the Catholic Church had in Mexico left an indelible impression on the culture and its people. By the 1880s and 1890s, the Roman Catholic Church had issued several decrees (May 19, 1886; December 15, 1886; July 27, 1892; August 3, 1897) banning the practice of cremation for Catholics, even excommunicating some long-time members who had chosen to be cremated upon their deaths.[114] For church officials, cremation violated a fundamental tenet of Catholicism: the resurrection of the body.[115] For example, on May 19, 1886, church officials, including most famously Pope Leo XIII, who branded cremation a "detestable abuse," decreed that burial was a "constant practice, consecrated by solemn rites of Church." Thus, the only method of bodily disposition for true Catholics was burial.[116] In 1908 church leaders continued to express their outrage over cremation by publishing articles in favor of burials. For example, in the English-language publication *The Catholic Encyclopedia*, church officials argued that burial was the truest form of piety, devoutness, and love. Simultaneously, cremation was "unseemly" since it destroyed the body, which was "once the living temple of God, the instrument of heavenly virtue, sanctified so often by the sacraments."[117] If an individual was a true Catholic who wanted to experience the rewards of the afterlife, there was only one option upon death: burial in a cemetery.

Don Porfirio believed the poor residents of Mexico City, whom he considered detrimental to progress, had only one choice: to accept their fate. Burials had become too expensive for the poor. Instead, cremation was a more useful method to improve public health and also continue the marginalization of the powerless. In his mind, if the city burned hundreds of corpses of the urban poor rather than burying them, the city could solve its health problems and finally be able to present itself as a modern capital equivalent to London, Paris, or Madrid. As early as 1891, Díaz had sent representatives from the Superior Sanitation Council to attend a panel on the hygienic benefits of cremation at the International Congress of Hygiene and Demography in London to explore the possibilities if cremation were adopted in Mexico. By the early twentieth century, more vocal support began to appear for cremation, especially from medical students. For example, José Najera authored a study on the advantages that cremation could offer President Díaz and his state officials. In Najera's opinion, cremation eliminated the dangerous odors, gases, and liquids that corpses emitted when buried.[118] Cremation was *the* simplest and most straightforward solution to the city's corpse problem.

In addition to the support cremation had received from state and medical officials, including the president of the country, a small but vocal group known as Episcopalian Methodists voiced their opinion about cremation in Mexico. Unlike the Roman Catholic majority who opposed cremation, especially of their own bodies, Episcopalian Methodists believed that cremation was the best option for disposing of the dead. For them, it was a disposition method that had resulted from the increasing presence of science in everyday life and thus an essential component for Mexico to become modern. According to Episcopalian Methodists, the Catholic press had routinely attacked science throughout its history, especially in the nineteenth century; they cited Darwinism as a theory that the church had ridiculed.[119] For this group, one of the church's most glaring shortcomings in its long history was its failure to embrace a relationship between science and religion. Despite the church's misconceptions, science was not out to destroy religion but instead to strengthen it.

Nevertheless, Episcopalian Methodists embraced cremation as the best disposition method available for a city struggling to improve public health. They saw cremation as an invaluable technological tool that they should promote and defend in the face of staunch Roman Catholic opposition. For example, shortly after the construction of the Marburg Crematory oven in Panteón Dolores, Episcopalian Methodists used their official newspaper, *El Abogado Cristiano Ilustrado* (The Enlightened Christian Lawyer), to publish columns dispelling the popular Roman Catholic belief that cremation violated an individual's capability for resurrection. According to one contributor, F. S. Borton, cremation did not prevent the possibility of resurrection because it was "the inhabitant and not the room that has value, the bird and not its old nest, the spiritual and not the physical."[120] Cremation, the paper also argued, was a process that mirrored the metamorphosis of a caterpillar to a butterfly.[121] The individual's spiritual body was more important than its corporeal one; thus, the technology surrounding bodily disposition was not seeking to attack religion but to present an alternate definition of piety for the modern world.

But for those who refuted this changing definition of piety, Episcopalian Methodists also provided evidence from scripture to support their view on cremation. The Bible, supporters argued, did not contain any passages that opposed cremation specifically. Here again, F. S. Borton argued that God had never envisioned the public health problems that plagued the modern world. Thus, man had to use the common sense God had given him to ameliorate public health problems. If this meant the use of cremation, then the Mexican people needed to put aside their "beautiful but false feelings and old but hurtful customs" and recognize cremation as a potential alternative to burial. In Borton's opinion, cremation was a more hygienic method for disposing of bodies since it reduced the body to ashes in hours rather than "by worms during six months, which produced an indescribable foul smell that harmed the living of the earth."[122] He even cited Genesis 3:19, "for you are dust, and to dust you shall return," as well as Ecclesiastes 3:20, "All go to one place. All are from the dust, and to the dust all return," to remind opponents that everyone turns into dust eventually, so why not accelerate the process if it meant improvements to public health? The sentiment expressed by Borton was one that cremationists around the world shared. Many continued to publish articles arguing that the release of gaseous matters into the atmosphere—the smoke from the crematory ovens—was not antichristian or antireligious; rather, it suggested the existence of another higher life form, a symbol of truly enlightened Christians.[123] To live in the modern world required individuals to reassess their beliefs and contemplate how they were going to deal with the fact that science, technology, and religion could all coexist.

For Porfirian state officials—many of whom maintained a religious identity but who were also generally secular in their approach to the deaths of others—the crematory ovens had become yet another example of a technology that could help them firmly control death and thus the bodies of citizens. The changing attitudes government officials had toward death signified that Mexican state officials were striving to achieve their version of modernity in spite of a major obstacle: ordinary Mexicans who would refuse to adopt new (*and forced*) technological changes.[124] For state officials and well-to-do residents, this technology for body disposition for the urban poor provided a quicker and more efficient method, a hallmark of Porfirian modernity. Ultimately, cremation allowed those in charge to erase the poor from the memories and landscape of modern Mexico City. The well-to-do remained the only group that continued to enjoy having their memories manifested physically in cemeteries, with mausoleums, sculptures, and well-decorated grave sites.

Conclusion

State officials remained committed to expanding their control over both the living and the dead well after Porfirio Díaz fled Mexico in 1911. After his exile, little changed for everyday people when it came to how the government sought to regulate their bodies in the name of modernity and public health. This trend continued throughout the twentieth century and into the twenty-first century; in November 2016 Mexico's minister of public health declared a public health emergency over the rise in the country's obesity and diabetes rates.[125] To combat this situation, in January 2014 state officials had enacted a 10-percent soda tax aimed at curbing the country's consumption of sugar, with proponents of the tax arguing that it had reduced the number of cases of "type 2 diabetes, stroke, heart attack and even death."[126] Yet, for all the good that it appears the government is trying to accomplish with such legislation, some opponents of the tax argue that not all of the evidence is in: soda sales continue to rise, and some consumers liken their consumption of soda to an addiction similar to that associated with nicotine.[127]

Nevertheless, the government remains committed to this reform program even in the face of myriad problems that continue to plague the country, such as drug trafficking, rising poverty levels, and the unexplained disappearance of political activists, including students, from the country. But the government's desire to modify and regulate the diet of its citizens in order to reduce mortality rates is not all that different than some of the desired goals from a century ago. While neither sugar nor obesity was at the forefront of the agenda in the early twentieth century, improvements in public health remained an integral component. Inventors continued to submit patents that highlighted how their inventions could help improve public health and allow state officials to exert control over the dead and the processes surrounding it in the name of technology and modernity.

One such example, submitted in 1925 by Spanish physician José Bassas Llados, outlined how his method of corpse conservation dovetailed with the desires of state officials to protect public health. He pointed out that conserving the dead was a method befitting only the most modern nations, an idea that resonated with state officials looking to justify changes in public health. "The cult of the dead," Bassas Llados wrote, "has throughout history been evidence of civilized and progressive people. Nobody can deny that the most appropriate way to honor the deceased is through the preservation of their body."[128] Accordingly, his conservation procedure would allow Mexicans to memorialize and offer their respect to the deceased. In particular, Bassas Llados lamented recent inventions that had tried too hard "to manipulate the body," removing organs or injecting bodies with questionable chemical solutions, crude methods that Bassas Llados felt "delivered bad effects for relatives." However, his preservation technique offered a solution. His method used "absolute efficiency, without subjecting the corpse to any operation or manipulation," as had been the case in previous years. In his opinion, "the body should play a completely passive role in treatment."[129] The procedure that Bassas Llados had developed involved covering the body in chemical solutions—made from thymol or formaldehyde—and placing it in a coffin. Once inside, an unnamed disinfectant gas would fill the air surrounding the corpse. Bassas Llados pointed out that after enveloping the corpse in a cloud of gas, the inside of the coffin would now be "completely sterile and therefore, no further decomposition will occur."[130] Such an approach presented state officials with an exclusive opportunity to a showcase modern and noninvasive method for preserving bodies, which would help protect public health in the capital.

Mexican state officials considered these promises to be attractive methods for achieving their goals. Both during and after the Porfiriato, many state officials continued to believe that to protect public health and highlight modernity in the capital, one of the most important issues that required their attention was how best to dispose of the dead. While cremation continued to be a viable option, it never surpassed traditional burial in popularity, largely due to the important role the body played in Mexican culture thanks to the centuries-long influence of Catholicism. Even new forms of funerary technology like that described by Bassas Llados failed to gain widespread popularity. Nevertheless, the tantalizing prospects offered by these technologies continued to attract the interest of state officials, who saw them as an opportunity to extend the reach of technology and reinforce population management techniques. Funerary technology was an essential part of the Mexican government's ongoing modernization campaign to control the lives of citizens by reducing the disorderly and chaotic methods surrounding how they disposed of and preserved their dead.

Yet this attempt to control how citizens chose to dispose of their dead often yielded unintended consequences. In particular, state officials failed to understand how members of the lower classes interacted with these new technologies. State officials' blind obsession with new forms of technology they believed would help create a modern Mexico yielded poor results. Investment in such technology only yielded the desired results if all citizens shared a similar ideological framework surrounding death. Members of the lower classes, however, rejected the state's new form of piety for one that better fit their worldview, one in which religion, not technology, remained an essential component of their lives. For those who did not believe in the official definition of modernity, which emphasized funerary technology as a method to improve public health and individual lives, this type of technology remained a useless tool that failed to improve their daily lives and reshape their understanding and acceptance of death. Tradition, not science nor technology, continued to represent how ordinary Mexicans chose to deal with death. For them, religion and its emissaries offered a more comforting outlook on the afterlife than any machine could. Neither science nor technology could save them, only God.

4

Undermining Progress

Workers, Citizens, and the Moral Economy of Death

Despite the introduction of new technology or free burials, along with a handful of other reforms aimed at improving public health in the capital, many of the city's lower-class residents chose to deal with death in a manner that was more convenient for them. While there were official rules regarding how to dispose of dead bodies, which ranged from a no-questions-asked free dropoff at local corpse deposits to registering the death with city officials who would subsequently issue death and burial certificates, officials failed to take into account the fact that in order to achieve their intended goal, their rules had to be followed. Introducing rules surrounding death was a far easier task than making sure all citizens adhered to them.

On March 15, 1906, for example, a police officer (badge number 583) watched from his post on the corner of Plaza Concepción and Zacate as a man hopped onto a tram that was out collecting corpses for delivery to Panteón Dolores. The driver, the officer later recalled, had asked the man for the corpse's papers, which he did not have with him. However, the lack of paperwork did not prevent the man from leaving the corpse on the tram; as the driver later told police, "he had forgotten the papers and ticket at his house" but promised to return with them. The driver, taking the man at his word, agreed to hold on to the corpse until the man returned. He never did. Eventually, the police sent the corpse to Hospital Juárez for examination, most likely to rule out contagious disease or potential homicide. There, medical officials determined that the deceased's name was Catalina Picaso; that name was scratched into the inside lid of the crude box she was left inside. According to the medical report, Catalina measured 160 centimeters in length (five feet, two inches) and was approximately thirty years of age, with black hair and black eyes, a small, wide nose, and a large mouth.[1] With the name and physical description of the deceased in hand, the police set out to gather more information.

Soon, an officer named José Alvarado told the investigators that a woman had come to him to tell him that she knew who had brought the corpse to the tram. The party responsible was the unnamed lover of Catalina, who according to the woman "was drunk and half-crazy with feelings," which had resulted in his behavior that had threatened public health. The unnamed woman was also able to provide the police with the address of Catalina's sister, Ausencia. But when the police arrived to question her, it was the wrong address, and as they noted in their report, "nobody knows where to find her." The investigation ended, and her final resting place remains a mystery.[2] However, it is quite probable that she ended up in one of two locations based on how city officials disposed of unclaimed corpses during this period. The first was an unmarked sixth-class grave in Panteón Dolores, alongside other corpses taken from local corpse deposits. The second was the cadaver department of the National School of Medicine, based on the institution's relationship with Hospital Juárez, where bodies were delivered as anatomical material to medical students.

Nevertheless, what remains important is the fact that incidents like this happened more frequently in the capital than officials liked. If it was not Catalina, it was the cadaver of a fetus wrapped in rags and placed in a box near an electric tram station or the bodies of unknown adult citizens abandoned near tram or railroad stations without any paperwork.[3] While rules concerning the proper way to dispose of the dead were printed on flyers and in newspapers, many lower-class citizens were unaware (due to their being illiterate), or if they knew the rules, they chose to ignore them. Tram drivers, usually from the lower classes, had explicit instructions to turn away anyone who failed to bring the proper paperwork (death certificate and burial ticket).

However, many chose to look the other way, especially if the grieving relative appeared distraught. Drivers often ended up having to defend their actions before police who were investigating the situation. In most statements, the drivers were either told by relatives that they had forgotten the paperwork and would return shortly, or people abandoned corpses on trams or near tram stops with the assumption that a decomposing corpse would prompt an immediate solution from the driver, who would most likely add the corpse to his regular collection and thus send it for burial in Panteón Dolores or another public cemetery. The disregard for the rules exhibited by relatives and even drivers was an important component of the moral economy surrounding death that lower-class citizens created during the Porfiriato. Their protest was not an outright rebellion against the government, which would soon follow during the Mexican Revolution, but more of a modest claim to have a say in the final destination of their corpses. While an official method regarding the proper way to handle death in modern Mexico existed, the government failed to take into account how the marginalized would react to that method. The individual interests of lower-class citizens took precedent over state officials' desire to establish a universal method for disposing of the dead.

The sight and smell of decomposing corpses abandoned on city streets—especially those of the urban poor—offended the sensibility of middle-class and elite residents. Moreover, their bodies represented an impediment to the progress Porfirian state officials had envisioned for the capital. While official records listed the number of corpses collected at cadaver deposits in the city for the year 1900 at 9,327, this excluded the numerous bodies found in the street, away from sanctioned deposits.[4] To solve the city's corpse problem in a more efficient manner, state officials and affiliated institutions would have to eradicate the urban poor's unseemly behavior and stress the importance of hygiene, especially when it came to disposition methods for the dead.

Porfirian state officials thought that "modern" Mexicans handled dead bodies according to the hygienic guidelines promoted by the government. This included placing corpses at official deposits (not in the streets) or making sure to meet corpse carts or electric trams at designated locations and times. Indeed, the majority of state officials viewed Mexico City as a laboratory, a place ready for experimentation with population management techniques and a modern public health system. Improving public health, however, especially as it related to the disposal of dead bodies, would be difficult. It meant eliminating certain behaviors and customs associated with people's everyday lives—especially among the lower classes, who state officials and well-to-do residents considered obstacles to achieving progress.[5] Many lower-class inhabitants ignored state regulations and instead chose to pursue strategies that fit their worldview and facilitated their survival in the city, especially when it related to death and determining the final resting place of a loved one.[6]

This chapter discusses how the urban poor created their own moral economy surrounding death to challenge the government's multilayered approach to regulating death and improving public health in the city. The Porfirian modernization project hinged on modifying the behavior and customs of the lower classes, particularly how they chose to live and die. By implementing and enforcing modern hygienic guidelines, state officials sought to erase lower-class behaviors and customs, which they believed were responsible for the city's deplorable public health conditions. However, altering their behavior would prove far more difficult than state officials had imagined, as lower-class citizens resisted the government's attempt to exploit their situation and chose instead to exert their own autonomy in the face of the modern state.

Tradition versus Modernity in Mexico City and Surrounding Cemeteries: Panteón Dolores, Magdalena, Tlalpan, Ixtapalapa, and Coyoacán

To modify the hygienic habits of citizens in Mexico City during the Porfiriato, especially those related to death, state officials relied on institutions like the Superior Sanitation Council to help spread the word about hygienic protocol throughout the capital.[7] But for the SSC to improve public health and hygiene, it needed to rely on citizen informants who kept its officers busy with information about the state of hygiene at their jobs or in their neighborhoods. The goal of the SSC was simple: help create an image of the capital that dovetailed with other changes in public health during the Porfiriato, all of which centered on proving that Mexico was healthy and modern.

Nevertheless, SSC officials often heard from concerned citizens about the patterns of behavior exhibited by their fellow citizens that they considered backward. For example, Panteón Dolores cemetery administrator Manuel Cervantes informed the SSC of a dangerous situation in his cemetery that had the potential to impact public health in the city. According to Cervantes, workers responsible for burying coffins had begun the practice of opening coffins before burying them in the ground in order "to assure the presence of the body."[8] The use of coffins among the lower classes, of which many of the workers belonged, was sparse. For their families, burials often consisted of a corpse wrapped in a shroud. Furthermore, some coffins contained valuable goods, like rings or watches, that the families wanted buried alongside the corpse for eternity. But these valuable were attractive to petty thieves, who sought to rob the dead at night. Even the nightwatchmen in Panteón Dolores were powerless to prevent robberies since the grounds were so expansive and the watchmen's numbers so few, which presented robbers with ample opportunities to profit. Items left at graves, such as marble statues, nickel-plated crosses, or metal trinkets, were easy to remove from the grave site and presented thieves will an opportunity to make a quick peso that nobody would miss.[9]

It appears that based on tradition or previous administrative rules, workers examined coffins to ensure that only one corpse, not two or three, was enclosed inside the coffin, as Cervantes said had occurred in the past. He admitted regretfully that the workers' behavior had become an unofficial "formality" under his supervision, lamenting the fact that repeated openings of coffins posed a potential threat to the city's public health.[10]

But the potential for additional income did not concern Cervantes. Instead, his fear was less tangible. Specifically, he believed that it was possible that if his workers continued to open coffins marked for burial, they would "spill" noxious odors and germs onto their clothes and into the air, which would have devastating consequences to public health. If this type of behavior had occurred at his cemetery, where he believed he had the ability to change the culture, then it was probably happening at other cemeteries throughout the city. Thus, he argued, it could lead to a potential pandemic, since these workers moved freely throughout the city. To combat such a situation, he proposed that physicians or their medical assistants could stamp or mark coffins in a way to ensure that workers would not open them.[11] But the temptation for poorly paid workers to earn additional money proved too much.

Nevertheless, Cervantes's proposal resonated with SSC members. Modifying the behavior of cemetery workers who represented a threat to public health was important. Before they could move forward, members suggested that they needed to hold a vote to determine whether or not to bring the issue to the attention of the federal government. The result was members voting in favor (ten to two) of bringing the suggestion before the government. In addition to Cervantes's suggestion about marking coffins, the SSC members added five new regulations for how best to prevent workers from opening coffins. First, they recommended that every coffin have transparent material made from glass or crystal above the corpse's face that measured "at least three inches long and two inches wide." This would provide "an unobstructed view into the interior to ascertain the presence of

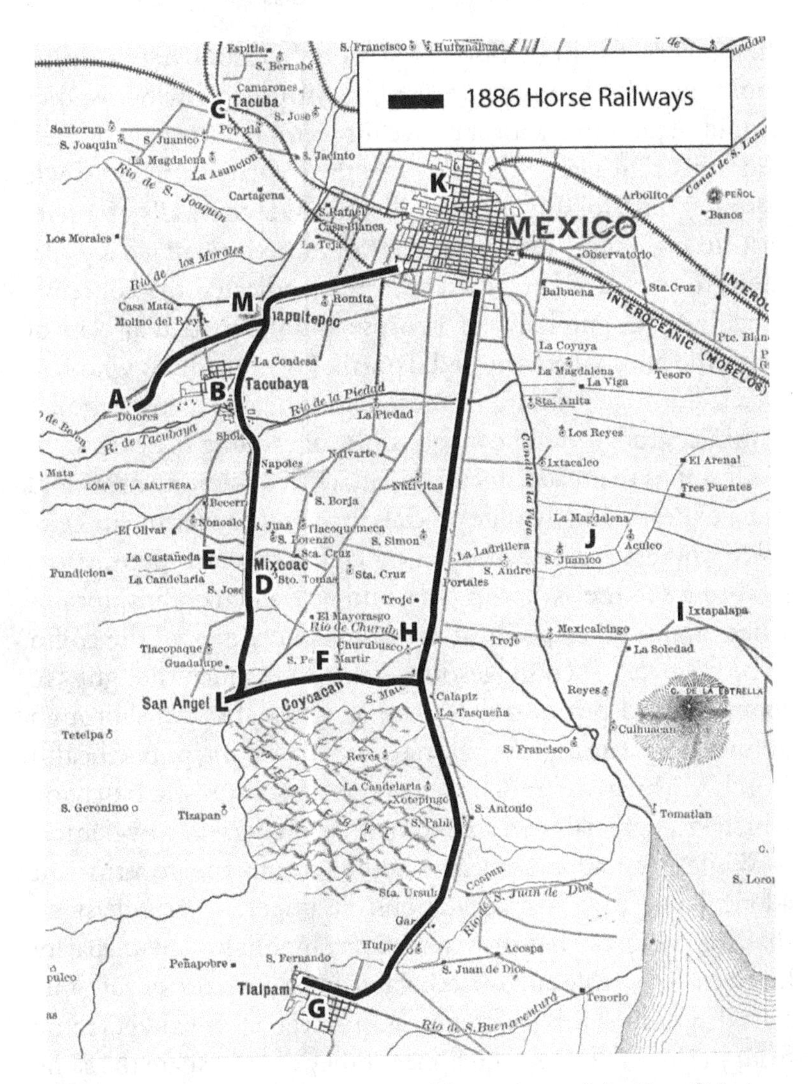

Map 3. General Carlos Pacheco, Map of the Environs of the City of Mexico, 1886. State officials believed that the railroad connecting Mexico City proper with myriad suburban towns would allow the government's modernization attempts in Mexico City to permeate throughout the valley of Mexico. Courtesy of Nettie Lee Benson Latin American Collection, University of Texas Libraries, University of Texas at Austin.

Key: **A.** Panteón Dolores, **B.** Tacubaya, **C.** Tacuba, **D.** Mixcoac, **E.** La Castañeda (Mental Asylum), **F.** Coyoacán, **G.** Tlalpan, **H.** Churubusco, **I.** Ixtapalapa, **J.** Magdalena, **K.** La Alameda Park, **L.** San Angel, **M.** Chapultepec Castle (Presidential Residence), the body that is to be buried." Second, in cases where families could not afford to provide a coffin with such a window, they should request that a local police inspector or "a person of their choosing" sign a letter verifying that the coffin contained a single corpse. Third, for anyone who died outside the walls of a state institution, a local funeral home would need to attach a document to the outside of the coffin that indicated the contents: specifically, the presence of a corpse and the particular class of grave the family had selected. Fourth, for individuals who had died inside state institutions (hospitals, jails, mental institutions), administrators would be responsible for sending the corpses in plain coffins that had a document attached to indicate name and cause of death. Finally, state officials threatened to terminate cemetery workers accused of opening coffins without permission. According to the SSC, workers could only open coffins in cases where a judicial or political authority had requested the coffin to be opened.[12] Such suggestions reveal how much thought SSC members had put into delivering a systematic and thorough approach for streamlining burials and protecting public health from the threat of cemetery workers' anachronistic behavior, which state officials considered neither modern nor hygienic.

While it remains unclear whether or not the government adopted these particular suggestions, it is important to understand that the incident that provoked them highlights the disparity between acceptable and unacceptable behaviors. Socioeconomic class determined this difference, as most cemetery workers came from the lower classes, which meant they did not share the same manners, customs, or tastes that elite citizens exhibited and wanted the lower classes to adopt. Their unfamiliarity with coffins or new technologies was a hallmark of lower-class citizens during the Porfirian era. Workers introduced to the wheelbarrow in parts of Mexico during the late nineteenth century, for example, chose to carry them on their heads rather than use them to move heavy or awkward objects as intended.[13] It had little to do with their intelligence but rather the value the technology held for them.

This type of behavior was routine. The cemetery was likely

the space where workers first encountered coffins on a consistent basis, their unfamiliarity with them prompting them to open the coffins to verify the presence of a corpse. The actions of cemetery workers threatened an imaginary social order that administrators and state officials believed they held over lowly state employees. To highlight the problems they had with workers, administrators often wrote to their superiors to inform them of the uncivilized behavior occurring at cemeteries: robbery, drinking, falling asleep on the job, and lateness to work were common behaviors they considered unprofessional. In fact, some administrators went to great lengths to report that these behaviors threatened the "order and progress" of the cemetery and represented an impediment to the continued growth of the capital.[14]

For workers unfamiliar with it, the coffin was a puzzling contraption for disposing of the dead, since traditional (and ancestral) methods of burial had involved various methods for body disposition, including wrapping the corpse in a shroud before burying it.[15] Additionally, some of the proposed rule changes meant that lower-class citizens would have to go to the police to receive confirmation of the presence of a corpse inside the coffin. The problem with this was the fact that the poor had at best a tenuous relationship with the police, since many police officers and administrators considered the poor to be thieves and criminals.[16] These suggestions reinforced the paternalistic attitude state officials had toward the urban poor. To modernize the city, government officials, including police officers, tried to force citizens to improve their daily habits, especially when it came to bodies, hygiene, and public health.

This forced change to individual behavior, especially something so private as hygiene or burials, was not met with enthusiasm or support from lower-class citizens. In fact, the changes state officials sought were largely ignored. Attitudes and customs surrounding both hygiene and death among the lower classes were not homogenous. Instead, much of how one chose to deal with these pressing issues depended on the individual's class, ethnicity, or geographic region. Nevertheless, the desire by state officials to define what was modern Mexican hygiene or death was not new. During the eighteenth century, for example, to reduce the potential for the spread of disease, state officials passed legislation to move burials from beneath churches to graves in cemeteries located far away from the majority of the population.[17] But what made the Porfirian example different was how death and public health intertwined with national identity as Porfirio Díaz pushed to control both elements to an unprecedented degree. To accomplish this goal, Díaz and his state officials created a patriotic death cult that celebrated important Mexican heroes like Benito Juárez (father of Mexico's liberal 1857 constitution) or Miguel Hidalgo (the father of Mexican independence). They also regulated popular activities surrounding the celebrations of the dead, such as a ban on the introduction of food and drinks inside cemeteries in 1900, which had been typical behavior for families visiting their dead relatives during the Day of the Dead festivities.[18] These regulations were essential components of Porfirian desire to establish an official relationship between state, citizen, and a modern death.

To accomplish this goal, Díaz and his officials "brought a cult of dead national heroes to a new pitch," beginning with Benito Juárez and expanding to include "principal politicians and intellectuals of the time," which even included rival politicians.[19] Not only did Díaz redirect how people would remember national heroes from then on, but he also began to manage and engineer how people celebrated and mourned in cemeteries, especially on holidays like the Days of the Dead (October 31 to November 2), when state officials and cemetery administrators believed violence and public drunkenness among the urban poor had become too large a stain on the progress state officials hoped to achieve. To remedy the situation, the government tried to increase the police presence in these cemeteries, mobilize volunteer watchmen, ban the consumption of food and drink inside the cemeteries, and close cemeteries before nightfall (at 6 p.m.), rather than the later hours that people had grown accustomed to.[20] "These acts of policing," as anthropologist Claudio Lomnitz has argued, "were not unambiguously directed at the poor, and could indeed be represented as subservient to the interests of the general public."[21] Thus, new regulations were integral to defining the relationship between state, citizen, and death in the modern world.

The government's attempt to regulate the celebration of the dead, especially during holidays, often backfired. Citizens continued to engage in banned activities such as drinking, aided by the fact that the pulquería across the street from Panteón Dolores remained open and thus a popular establishment for visitors. Instead of following new, modern rules for how to deal with death and dying in Mexico, lower-class citizens chose to ignore them.

Instrumental during this era to further cementing the unpopularity of such rule changes was the popularity of satiric *calaveras* (skeletons) printed throughout Mexico in penny press publications such as *El Hijo de Ahuizote* and *El Diablito Rojo*. Within the pages of these newspapers, countless skeleton images appeared to satirize and critique the hypocrisy surrounding mortality, public health, and modernity during the Porfirian presidency. In particular, the most lasting and well-known of these calavera prints were created by artist José Guadalupe Posada, whose calavera prints appeared in the pages of numerous penny press publications.

In "Choque de un carrofunebre," an electric tram has crashed into a funeral tram, causing a coffin to fall off the funeral tram and onto the track, breaking open and exposing passengers to the body of a well-dressed deceased man. While state officials sought to prevent well-to-do citizens from being the victims of progress, opting instead to promote how they would become the benefactors of such change, Posada's work expressed otherwise. For him, nobody was immune from the perils of progress and technology. In "Bicicletas," Posada's cyclists are all skeletons of different ages who are pedaling forward into the future toward progress. Nothing symbolized this advancement more at this time than the bicycle, a symbol associated with the middle class and its disposable income and leisure time to frolic carefree. However, one of Posada's cyclists has fallen, and his fellow cyclists continue to

Fig. 20. José Guadalupe Posada, *Choque de un eléctrico con un carro fúnebre impresion de originales en 1930* (collision between a trolley and a hearse). Courtesy of Museo Colección Andrés Blaisten.

pedal their bicycles over his body. But moving forward toward something, as Posada has drawn it, does not allow the others to escape death or the problems plaguing Mexico City. In fact, the idea that progress existed on some kind of linear path was false. The path to progress was more circular, much like a racetrack, just with no particular finish line in sight. Those who believed they were winning the race were mistaken. With each push of the pedals, the cyclists made their way closer to the truth: death was inescapable. Technology could not save them, just as it could not save their ancestors who had lauded the introduction of the steam engine. Whether they pedaled faster or slower, death was going to catch them sooner or later, just as it had the urban poor. For Posada, death was the great equalizer.

Nevertheless, this did not prevent Porfirian state officials from continuing to demand that citizens follow the official rules when it came to burials. The standing rule for all cemeteries within the Federal District was for cemetery administrators to turn away families who arrived without death certificates or burial tickets

FIG. 21. José Guadalupe Posada, Calavera "Las bicicletas." Dover Publications, scanned by Jack Child. Wikimedia Commons.

for the deceased. Paperwork was an essential component of the Porfirian plan to protect public health and inculcate inhabitants about proper burial methods since both measures reinforced the government's control the behavior of citizens. For state officials, once residents understood expectations, "order and progress" would soon pour through the streets of Mexico City and into the valley of Mexico, thus turning the country into the ideal picture of modernity that state officials desired.

But this was not reality. In areas away from the presence of federal and state officials, residents treated standards for burials differently. For those living in Magdalena, Tlalpan, Ixtapalapa, and Coyoacán, two important factors influenced whether or not individuals participated in the state's vision of modern death: cost and convenience. State officials from the Federal District set the price for burials. More prominent locations in the cemetery (first-, second-, or third-class graves) did not come without a hefty price, which reached upwards of fifty pesos ($1,900).[22] However, this price was out of reach for regular citizens. In Mexico, the average daily wage of low-skill workers at the turn of the twentieth century was one peso ($68). This meant that if the individual or his family desired a premium plot, it would cost roughly fifty pesos ($1,900), at minimum 8 percent of their annual income, which eliminated this option. Moreover, these prices did not reflect the true cost for the lower classes because wage laborers' work was far more inconsistent than salaried positions, and they tended not to work every day of the year.[23] The burial fees also required burial paperwork, which families could only collect at designated offices in the capital. So in addition to the actual expense of a burial, the family had to add the expense of not working for several hours, even half a day. This process made it difficult for working citizens in surrounding towns to take time off to travel to the city, a luxury that they could not afford. Working-class people were not in a position to leave work at their leisure. If they did, they risked losing their jobs, which meant losing money needed to feed their families. Even if the family took a train, their ability to arrive in the city depended on whether or not the railroads served their towns; if they did, the trip could still take anywhere from twenty to sixty minutes based on train schedules.[24] While the government knew this, officials did not change office hours to work with these citizens. Instead, they kept regular hours, so laborers could not leave for the city after work because the office would be closed already. As a result, life went on and individuals often chose to disregard the official regulations surrounding death and instead to participate in a moral economy of their creation.

In the town of Magdalena, just outside Mexico City—roughly six kilometers southeast of Panteón Dolores—one example illustrates how this moral economy worked to undermine the Porfirian state's desire to regulate death. On August 23, 1901, Judge Juan Z. Ceballos of Magdalena wrote to the state government in Mexico City to complain about an illegal burial that the cemetery administrator had told him about that had occurred at the local church cemetery in Magdalena. He asked for immediate guidance from state officials because, as he noted, this was "not the first time this has happened."[25] The church cemetery had stopped burying corpses several decades earlier, when state officials had created secular burial grounds throughout Mexico and made it illegal to bury anyone on church grounds or other unsanctioned spaces. The reason behind this shift was the fact that state officials believed on one hand that outlawing these burials would improve public health and on the other that it would afford the Mexican government more power over death.[26] Judge Ceballos informed state officials that he and the cemetery administrator were going to start an investigation into the illegal burial of the corpse of a local man named Bibiano Ballesteros. A week later, on August 30, the cemetery administrator told Judge Ceballos that he believed he had figured out which individuals were responsible for the illegal burial. During the course of the administrator's investigation, two names emerged: Pedro Abad and Angel Sánchez. With this lead, the administrator suggested that he have police officers question these two men to see what they could find out about the burial.

The police obtained some valuable information from the two men. Pedro Abad, the first man interviewed, was the assistant director of the cemetery in question, while Angel Sánchez was a neighbor of Bibiano Ballesteros. Abad confessed to the police that Sánchez had approached him "inquiring about how one could be buried in the church cemetery." He had told Sánchez that it was not possible to bury anyone in the church cemetery because this type of burial was no longer legal. Sánchez could petition the government of the Federal District for special permission, but he believed this was a wasted effort. But if Abad paid fifty pesos ($1,900) to the government for the burial, he could bury the corpse in the local public cemetery.[27] With Abad's story in hand, the police then questioned Sánchez about his discussion with Abad. According to Sánchez, he had never asked about burying Ballesteros because, as he put it, he was "neither a relative nor friend of the deceased" and thus had "no interest in the burial."[28] Faced with two conflicting stories, the police asked Sánchez if there was anyone who could verify his story.

Sánchez responded emphatically that he did have someone. He told the police that four witnesses, three of whom were relatives of the deceased, could clarify the situation. Toribio Ballesteros (Bibiano's brother) told police that the Ballesteros family, including himself, Perfecto Ballesteros (their other brother), Apolinario Ballesteros (his nephew, Bibiano's son), and an acquaintance named Vicente Mendoza had collected the funds necessary to pay for Bibiano's burial. However, after collecting the money from the family, the Ballesteros had changed their minds about the location of the burial. Rather than burying Bibiano in a public cemetery, as the four family members had intended, they now preferred that "the burial be done in the cemetery of the church of Magdalena" because it was a space that held religious significance to Bibiano and his family, despite the fact that the cemetery had stopped providing burials decades earlier. In the family's opinion, the public cemetery was insignificant to them.[29] As police continued to ask questions, the picture that emerged in their minds was that Pedro Abad and Angel Sánchez had hustled the family out of money and used their respective positions as assistant cemetery administrator and local "person of influence" to profit from the situation.[30] For the police, as agents of the state, illegal burials such as this were important cases to solve because the violations could be harmful not only to public health but to how state officials viewed the relationship between citizen and state. State officials had to trust citizens to follow established guidelines, and citizens had to trust that state officials held their best interests at heart and would protect them from potential threats.

However, the truth about the situation was more complicated than officials wanted to admit. For many of the people in the lower classes who lived at the margins of the capital, their past experiences had yielded little interference from the central government over how to bury the dead. It is true that state officials sought to replicate the rules and regulations surrounding death to these marginalized areas of the Federal District, but it was much harder to accomplish. In particular, state officials placed greater value on ensuring protection for public health inside city limits. For many residents outside city limits, as was the case with the town of Magdalena, Porfirian officials had kept a safe distance from prying into their daily lives, as they concentrated their efforts on improving public health in areas that were of greater importance. Thus, this distance helped to create a space that allowed suburban citizens to operate independent of state intervention. The fears, values, and habits of these citizens impacted how they chose to interact with death and the processes surrounding it. The result of such actions was the creation of a moral economy wherein lower-class citizens defined their own appropriate methods for dealing with death. This included ignoring prescribed rules in favor of burying the dead in places that held significance for families, even if it was illegal and required them to find willing participants in this economy who would accept cash in exchange for their services. The incident in Magdalena demonstrates how valuable these social exchanges were for the marginalized, who believed that the Porfirian government had violated their burial rights by forcing them to follow "modern" guidelines that moved burials away from sites that held significance (religious or otherwise) to a more secular and clinical final resting place in a public cemetery.[31]

Nevertheless, the police believed that Sánchez was culpable, having taken advantage of the family's desire for a meaningful burial. He had assured the family upon receiving payment that he would fill out all of the required paperwork and take care of all other potential matters surrounding the burial. As a result, Sánchez split the fifty pesos ($1,790) with Abad, since he had used his position as assistant caretaker to prevent the discovery of the event, or so he thought.[32]

When the case closed on September 30, however, the judge assigned to the case issued two rulings. First, he ordered a team of legal medicine physicians (forensic doctors) to exhume the body of Bibiano Ballesteros from the closed cemetery and move him to the public, secular cemetery of San Francisco, as was the rule for the town of Magdalena. Second, he issued a verbal reprimand of cemetery administrator Ceballos for knowing of other illegal burials but choosing "to do nothing about it." As for Sánchez and Abad, the judge issued an unspecified punishment in his ruling.[33] Despite the absence of detail surrounding the punishment, the case itself remains significant for understanding how the moral economy surrounding death operated at the margins of society. The desire to bury a family member in a meaningful location demonstrated to what degree individuals were willing to go to have their or their family members' desires met. Rather than follow the official rules concerning a proper burial, citizens who lived outside Mexico City limits used their distance from the reach of the government to carve out an ideological framework that made sense to them. As a result, this often meant circumventing official protocol and instead continuing to engage with decades-, even centuries-old traditions when it came to death.

Another incident that illustrates how this moral economy had become the status quo surrounding death for lower-class citizens residing outside city limits occurred in Tlalpan. Several complaints from local residents regarding how some families disposed of dead bodies at the local cemetery appeared on the desk of Federal District Governor Ramón Corral.[34] On April 12, 1902, Corral requested that the Civil Registry—the office that oversaw birth and deaths—come to Tlalpan to conduct a review of the condition of the city's public cemetery. Corral sent Agustín M. Cordero of the Civil Registry to investigate the complaints. His findings provided a bleak picture of the cemetery's condition, one that would require immediate action if things were to improve.

Cordero's report detailed why local residents were upset with what was happening at the cemetery. Since the cemetery was full, with no additional burial space available, families had chosen to prop corpses up against outside cemetery walls. Moreover, Cordero found that in the southeast corner of the cemetery, the walls had "cracked and collapsed," which worried him since he believed it was quite possible that families saw this is an opportunity to bury their dead for free.[35] The report also mentioned that the shed cemetery workers stored corpses in was missing both a wall and roof. Thus, the corpses remained exposed to the elements, which could have dire consequences to the state of public health in Tlalpan. Despite being outside the immediate purview of state officials in the capital, such conditions were still worrisome. In particular, Tlalpan was one of several suburban towns (including Coyoacán and Churubusco) that received daily train service from the capital. Such fluidity between the capital and suburb meant that more people risked exposure to public health threats from the cemetery as well as an unflattering impression of the power and influence state officials held in suburban towns.

To ameliorate the situation, Governor Corral asked Cordero to provide him with an exact list of what he believed required repair as well as an estimated total cost. On April 18 Cordero submitted his list to Corral, which included the construction of a new wall in the southeast corner, repairs to the corpse shed, and a new house for the current cemetery caretaker, who could keep close watch on the corpse problems, all of which brought the total to 515 pesos ($5,010).[36] Ten days later, Ramón Corral informed Cordero that he would provide all of the money he had estimated for repairs.[37] Cordero thanked the governor but also took the opportunity to propose that the governor issue an exhumation order for corpses buried the longest in the cemetery to create more room for present and future burials. Corral agreed with Cordero's proposal, and on May 15, he announced that exhumations in Tlalpan Cemetery would begin shortly and that families with loved ones buried in exhumable areas needed to come to the cemetery to collect the remains, which had turned to bone and thus posed significantly less of a threat to public health.[38] Families had five days to collect the remains for possible reburial at another cemetery, but if they failed to collect in time, the bones would end up in the ossuary, where they would likely become study material for medical students or medical schools.[39] Corral's edict attempted to make it clear to the residents of Tlalpan that when it came to death, the official method was the only choice. Perhaps the threat initiated by state officials to give away the bones of the deceased motivated families to collect the remains so that they would not end up in the hands of students.

State officials had invested heavily (financially, emotionally, and ideologically) in Mexico City cemeteries throughout the nineteenth century. In particular, cemetery exhumations helped to support population management techniques and reinforce the desire to control the bodies of citizens.[40] By the early twentieth century, however, the Porfirian state's investment in improving public health, which included monitoring conditions at cemeteries, had begun to include nearby towns and villages, especially those connected by rail. Nevertheless, this attempt to exert control over death prompted people living in these areas to react in ways that state officials had not considered. These citizens considered their traditions and way of life under attack by state bureaucrats who sought to impose a distinct vision of public health and modern burials. These changes violated the moral economy surrounding death that these individuals had created over decades, even centuries. They chose to create their own definitions of public health and modern death from their own experiences, social mores, and morals, which often clashed with how state officials had defined public health and death.

Thus, a significant number of individuals living on the periphery of Mexico City ignored the official rules surrounding death. In particular, the relationship lower-class citizens had with death was far more impactful on an individual's daily life than mandates from state officials. One such example of this conflict occurred on January 28, 1903, in Ixtapalapa, a mostly indigenous town southeast of Mexico City whose residents lived in abject poverty.[41] The Superior Sanitation Council received a letter detailing how municipal cemetery workers in Ixtapalapa had dealt with exhumations. In the letter, brothers Ramón and Tomás Cedillo, along with neighbor Tomás Luna, explained that for several days workers had been exhuming corpses that "had reached their temporarility."[42] This meant that, when originally buried, the families of the deceased had paid for a five-year term burial rather than a burial for perpetuity, largely because it was more affordable for short-term burials. According to the Cedillos and Luna, despite five years having passed, the exhumed corpses had not decomposed completely. However, this had not prevented workers from removing corpses and taking them to a cave on cemetery property, where they left the corpses "out in the open air, producing unbearable miasmas" for those living nearby.[43] The situation at the cemetery would cause the SSC to investigate.

After receiving the complaint, physician and president of the Superior Sanitation Council Eduardo Licéaga sent José Ramón, the secretary general of the SSC, to Ixtapalapa to investigate the complaint. According to Ramón's report, which he presented to SSC members on February 7, the situation in Ixtapalapa was alarming. Not only did the "extracted bodies give off an unbearable stench" that permeated the air, but some of the exhumed corpses had "distended bellies containing masses of organic material that give a repugnant aspect to the remains." This represented a potential public health nightmare that could blow up (literally) in the faces of state officials. Furthermore, it gave critics additional evidence to support claims that the picture of modern death painted by state officials was not as it seemed. To improve the situation, José Ramón suggested that the SSC order the cemetery to address the situation in two ways. First, the cemetery needed to bury all exhumed corpses immediately, regardless of the state of decomposition. Second and most importantly, the cemetery had to implement the updated rules found at Panteón Dolores to extend temporary burial terms by two years, which brought the term to seven years. Ramón believed the additional two years would reduce the corpse entirely to bone, and thus there would no longer be a risk of exhuming bodies partially decomposed.[44]

SSC members discussed José Ramón's proposal and returned with a resolution four days later on February 11. While the members agreed with Ramón's assertion that there was "a bad odor" at the Ixtapalapa Cemetery, they believed that exhuming corpses after five years was unacceptable. According to the discussion, the exhumations at Ixtapalapa corresponded to official protocol. As they pointed out, article 239 of the 1877 Sanitary Code stated that the Superior Sanitation Council "will set for each cemetery the time that human remains stay in the grave and outside this term, the only way to permit exhumation was a direct order from the government of the Federal District."[45] If the SSC had said five years, then five years was enough time for exhumations. Nevertheless, the exhumations in Ixtapalapa—the first carried out at the cemetery—were of corpses buried in 1895 and 1896. Their partial decomposition, according to the SSC, had less to do with the term and more to do with the "special conditions of the terrain" caused by inadequate rainfall.[46] In spite of the outrage expressed by Ixtapalapa residents, the SSC decided that the actions of cemetery workers did not warrant any further attention. While there was no official reason as to why it was not an urgent matter, there remains a simple explanation: SSC members lived in Mexico City, not in Ixtapalapa. Therefore, the potential public health problems that residents believed to exist did not affect the day-to-day lives of SSC members or other residents in the capital. The threat was real, but it was one that existed on the margins, not at the center of the area where the SSC operated.

The threat that this behavior posed to Porfirian "order and progress" went beyond Ixtapalapa. On June 5, 1903, in Panteón General in Coyoacán, a suburban town much closer to the capital, the family of a deceased little girl named María Jesús Salinas Labastida noticed something strange at her third-class grave site. At the grave the family had found fresh dirt, which puzzled her father, Crescencio Salinas, who told authorities that on a previous visit the grave site "had been normal."[47] Angered by the situation, Salinas went to the office of the cemetery manager, hoping to get an answer.

The cemetery manager provided Salinas with a brutally honest response to his inquiry. The cemetery, Salinas later informed SSC officials, did not maintain detailed records about burials, an act antithetical to the modern image state officials wanted for cemeteries. The manager admitted that since "the graves have no numbering or signs, it is quite difficult to know which of them are filled and which are not." Therefore, without a comprehensive index of the deceased (names, burial dates, or grave numbers), the manager had selected María's grave for exhumation unknowingly. According to Salinas, the manager told him that he believed that "the deposited corpse had already served its term" and thus he had had her body exhumed on May 30 to make room for another corpse.[48] To prepare the grave, the manager had asked cemetery workers to empty it by removing the existing coffin. However, once workers lifted the coffin out of the grave, they saw "the state of the coffin and noticed the stench that it emitted" and told the manager, who realized he had made a mistake. The exhumed body was not of someone who had died years ago but rather a more recent burial, which meant that it had not decomposed entirely. The manager decided to have the workers put the coffin back into the grave and simply add the new coffin as well, since it was much easier to stack the coffins on top of each other than to find a new grave site.

Despite the honest mistake and explanation, an angry Crescencio Salinas contacted Ramón Noriega, a member of the Superior Sanitation Council, to ask for an investigation into the practices occurring at Coyoacán Cemetery. Details remain meager. Nevertheless, one piece of information remains crucial to the historical record: the accidental exhumation was not the first time something like this had occurred. When Noriega arrived in Coyoacán, he noted that several local families had come to him to express their concern that the manager had given "little care and bad treatment" to the dead at the Panteón General.[49] While there were no more details offered, the fact that families sought out Noriega illustrates the disorderly and chaotic nature of burials that took place on the periphery of the capital. Regulations surrounding burials extended beyond the boundary of the capital to all of the Federal District; however, distance from regulatory agencies such as the SSC helped to foster an environment wherein official protocol held a tenuous position over both citizens and workers.

Incidents of this sort in Magdalena, Tlalpan, Coyoacán, and Ixtapalapa, all towns on the periphery, suggest that state officials frequently had to compete with the autonomy that individuals expressed when it came to burials and public health at the margins. State officials expected that all institutions involving public health, which included cemeteries, would adopt the official approach to public health. The use of statistics, proper hygiene, and handling of the dead, in the opinion of Porfirian state officials, demonstrated that progress was possible in Mexico City.[50] Yet cemeteries became contested spaces where the goal of creating a hygienic environment often failed as workers and administrators both demonstrated a lack of concern for public health and safety by ignoring the prescribed rules. The majority of inhabitants living in these suburban towns were members of the lower classes who fiercely preferred to advance their independence by creating a moral economy surrounding death that had its own set of unwritten but nonetheless official rules of the people.[51] Thus, city rules held little value in the suburban environment where lower-class residents lived. They often chose to ignore the wishes of the Porfirian state to illustrate the point that they did not believe that state officials knew them or their problems. As a result, the people were left to take care of themselves and their problems in ways that fit their lives.

A Threat to Public Health: People and the Rules Surrounding Death

The moral economy that existed around the issue of death was not just limited to residents in the outskirts of the capital. Challenges to the Porfirian state over who was the adjudicator of death still occurred inside city limits. One such incident happened on June 2, 1902, when six-month-old Eugenio Lugo stopped breathing. His parents tried to revive him, but after several attempts, they abandoned their efforts and rushed the baby to a neighborhood medical clinic, where they hoped he could be resuscitated. Unfortunately, he could not, and the clinic pronounced him dead on arrival. In between the shock and tears, the sobbing parents asked the attending physicians if it were possible for them to take their son's body home so they could collect some of his toys and dress him in his best clothing before burying him. The parents promised that they would only be gone an hour or two, and upon their return, they would collect the death certificate required by law to register the burial, pay the associated fees, and retrieve the burial ticket that provided entrance to the public cemetery.

An hour or two soon became six, and neither parent had returned with the infant's body. The physician who had agreed to allow the parents to leave began to panic. Against state regulations, he had allowed them to travel with a decomposing corpse, which represented a tremendous threat to public health, one that could cost him his job. To rectify the situation, the physician contacted the Superior Sanitation Council to tell the officials about the mistake he had made in the hope that their agents could find the family before it became even more of a health problem.[52] After all, he did not want to be responsible for any potential damage to public health that a decomposing corpse could cause since all liability regarding the release would be placed on him.

The SSC chose agent Luis Parano to investigate the situation and tasked him with finding the body before it was too late. In a stroke of luck, Parano was able to obtain an address from the intake paperwork that the parents had filled out at the clinic. With the address in hand, Parano began his search for the missing corpse.[53] Searching among a row of dilapidated houses, Parano was able to match the address he had found on the paperwork: house number two on La Tercera Calle de las Artes. Standing outside, he knocked on a rotting wooden door, and the wood splintered around his fist. Several minutes passed without an answer. Eventually, the door opened, and an old woman yelled at Parano to go away, she was not interested in whatever he was selling. He quickly corrected the woman, telling her that he was from the Superior Sanitation Council and that he was looking for a corpse, and thus, she needed "to hand over the infant's body" before it caused any more problems. Upon hearing this, the woman opened the door all the way, Parano wrote in his report, telling him that she had no idea what he was talking about but she could assure him there was no corpse in her house. Nevertheless, Parano continued to pester her, hoping to extract the truth. He told her that he knew she had gone to the clinic earlier that day with the corpse of Eugenio, and she needed to return his body to make things easier on her, before the authorities got involved. Again, she stuck to her story and told Parano she had no idea what he was talking about. Perhaps he had the wrong address, the women told him, because she admitted she was far too old to have an infant son.[54]

Convinced that she was not who he was looking for, Parano returned to SSC headquarters emptyhanded. Upon his arrival, there was more information waiting for him regarding the missing boy's case. According to a medical assistant from the clinic, someone at the clinic had just sent the death certificate for Eugenio to burial officials in Huichapan, a town in the state of Hidalgo, almost two hundred kilometers away from Mexico City. Parano quickly informed his superiors, and the SSC telegraphed all of the town councils between Mexico City and Huichapan to ask them to be on the lookout for a man and woman trying to obtain a burial ticket for a dead infant named Eugenio. However, the SSC received the same message from all the town officials: no one matching that description had tried to obtain a burial ticket for a dead infant that day.[55]

But something was not right. How could the parents have gotten to Huichapan so quickly? According to Parano's report, he went back to the house on La Tercera Calle, where the unnamed old woman had turned him away, because he believed she knew more about the situation than she had admitted. Upon returning to the house, he asked the woman how she knew Eugenio's parents. It was not a coincidence, he told her, that the family would list some random address in the city, unless they knew her in some capacity. Whether worn down by guilt, tiredness, or something else, the old woman confessed to Parano that she indeed knew the parents through a friend of a friend. While she had lied about knowing them, she swore to Parano that she honestly had no idea where the parents had gone. They had asked her if they could put her address on some paper they had to sign, but she had no idea what type of paper. While she regretted not telling Parano the truth to begin with, she informed him that she did have something valuable that he was after: the location of the boy's corpse.

Next, the woman took him to a few streets over to La Quinta Calle de las Artes. There, according to Parano, she pointed to a small, crudely constructed, and unnumbered shack (*jacal*) that her friends Ignacia Hernández and José Martínez—who were friends of the Lugo family—had rented. He thanked the woman for the truth as she turned and left to return to her home. As Parano knocked on the door, he noticed that no one was there. So he waited. A few hours later, Hernández and Martínez returned, and immediately, Parano questioned them about their relationship to the Lugos and the missing corpse of Eugenio. Within minutes, the couple confessed that they were friends of the Lugos, who had asked them to "hold onto the body for a day or two."[56] Hernández and Martínez admitted to Parano that they had agreed to keep the corpse because the couple promised to return shortly and that they believed the sudden death of the boy had caused the couple great sadness. The couple assured Parano they regretted taking the corpse but that they had no idea officials had issued a death certificate or that it was a crime for them to hold on to the corpse.[57] An interesting piece of the report that sheds light on the lower-class intimacy with death was the fact that they never once mentioned their health being at risk by their harboring a decomposing corpse in their home. Such an omission supports anecdotal evidence found in popular accounts such as Oscar Lewis's *The Children of Sánchez* that it was common for lower-class citizens to hold onto corpses for several days in order for the family to celebrate and mourn the deceased.

It remains unclear whether or not the SSC ever found the Lugos. But what is more important was the fact that the incident illustrated the type of behavior surrounding death that state officials were trying to eliminate. Death was something that the

Porfirian state wanted to remove from sight. The creation of a prescription for death, how to handle and bury the body and, in some cases, how to mourn it, was a powerful act on the part of state officials. As archeologist Timothy Taylor has noted, the attempt to deny bereavement reinforced obedience to the state and created a controlled zone according to the desires of state officials.[58] Instead of following the state's prescribed rules for death, which included how to handle dead bodies, the urban poor chose to deal with the situation in ways that were familiar to them. Death was a separate category like crime or prostitution, subject to the "advanced technical standards and new moral sciences" of modern Mexico.[59] To reach a level of true modernization, Porfirian state officials needed all citizens to follow the precise measures outlined in the state's agenda for how to establish a hygienic and orderly public health environment.

However, traditional customs honoring the dead remained prevalent, especially among lower-class citizens. The *velorio*, where families laid the deceased on a table inside the home, "surrounding him or her with candles, and inviting kin, friends, and acquaintances to observe the body and pay their respects to its soul over the course of a few days," stood in direct contrast to the state's desire to make death a controlled, impersonal event.[60] The introduction of a systematic approach to death, highlighted by the need for a death certificate and burial ticket and specific rules regarding where, when, and for how long families could bury corpses were all options that state officials implemented in an attempt to define what it meant to be modern. Nevertheless, these options often clashed with how the average Mexican sought to deal with death. For the urban poor, resisting the state's clinical approach to death was not a premeditated act but rather as an instinctual response to the situation at hand, where living close to the margins offered little alternative. For them, the death of a loved one was one of the last vestiges of a Mexico wherein individuals could exert some autonomy over their behavior.

One such example happened on January 29, 1904, when Benjamín Pérez arrived at the local police station in Tacuba, a suburban town four miles northwest of Mexico City. Tacuba was a town with a rich indigenous past, like many central Mexican towns. It was the home of the Tepanec Indians with a rich history dating back to the Aztec Confederation.[61] After the Spanish conquest of central Mexico, the Catholic Church set out to convert the indigenous populations to Catholicism. However, Catholicism did not take exclusive hold over individuals' religion, but rather a combination of Catholicism and indigenous religion predominated. During major Catholic celebrations such as Holy Week, the church attempted to mold individual belief systems, especially pertaining to death, which visitors characterized as full of "superstition, quasi-solemnity, much noise and tawdry display."[62] This had to influence how Tacubans chose to deal with death given the reverence and celebration showcased by the church versus the more distant and matter-of-fact view of death held by secular state officials.

At the police station, Pérez informed the officers about a potential threat to public health near his home. According to the police report, Pérez found a corpse while collecting firewood near his house. As he had told officers, when he bent over to pick up some firewood, he had noticed "a bundle of rags" beneath. Curious, Pérez pulled on the rags, which started to loosen, uncovering the decomposing head of a semiburied infant.[63] While the police believed the story seemed far-fetched, they decided to check it out anyway to rule out the potential public health threat and possible homicide.

To investigate the situation further, the police chief sent his legal medicine team, medical physicians Pedro Alfaro and Emilio Pineda, to the scene with Pérez. At the scene, physicians encountered something more gruesome than what Pérez had described to police. As the physicians located and collected the infant's head, still wrapped in rags, they wrote that something (or someone) had separated the head from the body crudely. While searching the area for the rest of the infant's body, they heard a strange noise coming from the woods nearby. They decided to investigate, and as they got closer, the physicians noticed a pack of feral dogs fighting over something. They discovered that the dogs were chewing on the remains of the infant in question. In order to determine cause of death and to rule out homicide, the physicians noted, they had to retrieve the remains. With no other choice, they charged at the pack of dogs. As they ran at the dogs, the remains of the infant fell to the ground, allowing the physicians to collect them quickly.[64] With the remains now in hand, the doctors headed back to their laboratory in the police precinct to conduct a thorough examination into the exact cause of death and whether or not the corpse had posed a serious threat to public health.

However, this would prove difficult to ascertain for the doctors. Back at the lab, doctors Alfaro and Pineda discovered that the feral dogs had destroyed too much of the infant's remains, as it had no intestines or legs left. The only remaining parts in a condition to examine were the head, torso, and two arms. As the autopsy report explained, without an entire body, the cause of death was impossible to determine. Yet in their medical opinion, the one thing they knew for sure was that the infant had "not lived or breathed outside the womb": it had been stillborn.[65] Thus, the corpse presented no threat to public health presented.

While no public health threat existed, it did not prevent authorities from uncovering additional pertinent information. The outcome of the case disappeared from the historical record, but the incident remains important because it provides a unique window into understanding how regular citizens chose to interact with an environment that to them continued to be more stringent by the day. According to the field report submitted by the physicians, someone, presumably the parents, had tried to bury the infant in a small hole that measured 30 centimeters (11.8 inches) in length by 25 centimeters (9.8 inches) deep. But more important than the size of the hole was the location of the attempted burial. The doctors had found the corpse 65 feet (19.8 meters) from the walls of a Catholic chapel known as Santiago Huixachuac.[66] Such a burial location was not accidental. The approach to death emphasized by Porfirian officials sought to place great value on following specific hygienic rules when it came to dead bodies. However, for many lower-class citizens, this clinical gaze clashed with their religion, much of which emphasized traditional Catholic rituals surrounding death, in which burial near a church or a church graveyard was tantamount to a restful afterlife. But beyond the spiritual aspect, there were more practical reasons why an illegal burial became a possible option for families: cost and worldview. For lower-class citizens, adhering to the rules of hygiene or medical science elites had envisioned for Mexico required additional time and paperwork, both of which placed an undue burden on the poor.

Death in the twentieth century had lost that intimate emotion it had held in previous years. For state officials who applauded the benefits that science brought to society and that influenced their population management techniques, death had become something to calculate, a clinical and emotionless statistic. As historian Charles Hale has pointed out, the philosophical ideas guiding Mexican state officials at the time was often referred to as positivism, a theory of knowledge that promoted that the scientific method was man's only means of becoming knowledgeable. In the past, this knowledge could only occur with the assistance of the Catholic Church and its priests, whom Porfirian state officials, physicians, and hygienists had eliminated and replaced with themselves. Thus, officials would observe, experiment, and offer solutions to societal problems based on a system that valued research, organization, and administration, an approach that mirrored what took place in a laboratory.[67] For state officials, science had become the only religion, which made physicians more valuable as they held the answers to society's questions, just as priests had in an earlier era.

Clashes over Death: Workers, Medical Schools, and Hospitals

The struggle between the different approaches to death also appeared within public institutions, where state officials tried to make sure that the official view on death trickled down to everyone working there. While the state officials who filled out high-ranking positions in these institutions considered themselves to be influenced by secular and modern scientific principles, the individuals employed to handle bodies in each institution failed to share the same outlook on death as their bosses. The intimacy with death in the lower classes influenced how they approached handling bodies, especially when economic opportunities existed.

One example that illustrates this clash between lower-class working culture and institutional culture occurred at the prestigious state-funded National School of Medicine. On August 31, 1905, physician Eduardo Licéaga—the director of the National School of Medicine and president of the Superior Sanitation Council—received a shocking letter from a unnamed individual who accused one of the school's medical attendants of selling cadavers to students as well as providing them with additional opportunities outside of class to practice dissection on campus.[68] Such activity angered Licéaga. For medical attendants, the opportunity to be on the precipice of such integral work contributing to the betterment of young minds and improvement of public health should have been enough of a reward. The money made by selling corpses to students undermined this value. Additionally, creating a controlled, hygienic, and safe working environment at the university was important not only for public health but for furthering the population management techniques of the Díaz administration, which placed great emphasis on following the rules. Knowingly violating these rules, however, put everyone at the university at risk in the opinion of state officials. As models of progress, professors and other employees had to follow a hygienic protocol surrounding the movement and disposal of dead bodies in order to protect public health, which was part of the Mexican state's modernity campaign.[69] To put a stop to the potential devastation that an incident of this magnitude could cause, Licéaga turned to the university's provost, Dr. Eduardo Vargas. Vargas would spearhead the investigation to determine whether or not Guadalupe Rodríguez, a medical attendant in the school's Topographical Anatomy Department, was the party responsible.

To get to the bottom of the accusation, Vargas decided it was best to spend a week studying how the school received and distributed cadavers. Each day was the same, Vargas wrote: employees from Hospital General, a public hospital, drove the cadavers of unclaimed patients to the school. There, they delivered the cadavers to an on-duty medical attendant whose job was to record and subsequently track the distribution of cadavers. When Hospital General employees offloaded cadavers, a medical school attendant marked the skin of each cadaver with a letter of the alphabet. For example, the first cadaver received an A, the second a B, and so on. According to Vargas, the school had introduced this method to make the distribution and tracking of cadavers far more systematic than the previous method, which had relied on individuals recording the information in logs. After recording which letter of the alphabet each cadaver received, the attendant would then deliver the corpses to the assigned departments. Once delivered, the medical attendant would prepare the cadavers for dissection. After the class had finished dissecting the cadaver, the attendant would collect what remained of the corpse and place the body in an open-air storage room, where it stayed until the next day (or two) when dissection resumed.[70] With a solid foundation for cadaver distribution and collection, it puzzled Vargas as to how someone could even sell cadavers without arousing suspicion. To find the underlying cause of the situation, he decided to talk to the alleged recipients of the cadavers: the medical students.

According to Vargas, most of the students he interviewed were unaware that any of the attendants had sold cadavers to students. Nevertheless, as he interviewed more students, he began to find evidence that supported the accusation against Guadalupe Rodríguez. Several students from the topographical anatomy class informed Vargas that Guadalupe Rodríguez was their attendant and that at the end of the class, he often failed to return the cadavers to storage, which was against departmental protocol. Instead, the cadavers "remained in the class without destination." In Vargas's opinion, this unsupervised time presented a prime opportunity for Rodríguez to sell cadavers to students interested in increasing the time they could spend practicing dissection independently. The possibility of this was all the information that Vargas needed before concluding that the accusations against Rodríguez were true. Thus, Vargas informed Licéaga that Rodríguez was the culprit responsible for selling cadavers clandestinely. Immediately, Licéaga fired Rodríguez for "distributing cadavers to students for tips."[71] Such behavior was antithetical to the environment and culture that Licéaga had worked hard to promote over the years. He expected workers to behave in ways that promoted the good work the university had done and that reflected the positive elements of Porfirian modernity.

Additionally, the Vargas investigation had also uncovered the reason why cadaver trafficking had even occurred at the university in the first place. For Licéaga, there were ample opportunities for students to dissect cadavers in class. But he had failed to take into account that before his term as director, and even during his early years as director, a medical student culture existed that prided itself on students' paying for bones, skeletons, and cadavers, which was presumably different than what he had experienced during his own medical training. A student's ability to operate outside official channels was a rite of passage that students believed was "necessary and indispensable for their studies."[72] Learning to fend for themselves was useful for their future careers, especially if assigned to more rural areas in Mexico far from the watch and support of the Mexican government. While Licéaga believed he had steered the university in a modern direction, the cultural legacy of cadaver trafficking at the school undermined this idea and demonstrated how students sought to define their roles in this system.

For students, untapped potential existed in the symbiotic relationships they built with medical attendants like Guadalupe Rodríguez. Medical assistants earned additional money while students earned additional studying time, and thus, together they created a system of reciprocity that was good practice for their futures. Meanwhile, for medical assistants, the relationship was a practical one, since it allowed them to augment their standard incomes by allowing students to purchase cadavers that at some point would be disposed of anyway. For example, the monthly salary of a medical assistant in 1905 was only ten pesos ($601), which was a low amount considering the macabre nature of their work.[73] The risk posed by participating in such an illegal activity often outweighed the alternative, doing nothing, which did not help pay the bills. As a marginalized group, these workers chose to violate the rules in ways that provided immediate and often short-term relief to them and their families rather than continuing to suffer economically, even if it meant undermining the Porfirian state and the rise of unemployment.

As a staunch supporter of the Porfirian administration and its goals, Eduardo Licéaga used the Guadalupe Rodríguez incident as a way to change how he and the National School of Medicine would distribute cadavers to various departments going forward. The new method emphasized the importance of maintaining a paper trail that would eliminate any potential liability stemming from illegal activities. The new method required medical assistants to provide documentation to both individual department heads as well as the secretary of the university that detailed the cadaver's name, assigned letter (A, B, and so on), and date of delivery. Additionally, once dissection of a cadaver began, the attendant recorded the name of the receiving class and department. Someone, usually the attendant, also had to provide a list of what parts of the body the class had dissected. This way, any additional incisions discovered would indicate unauthorized activity outside of class.[74] This approach to cadaver distribution made it much easier for university officials to identify illegal activity. More importantly, it allowed Licéaga, as an agent of the Porfirian state, to try to exert total institutional control, which was a similar approach that President Díaz and his state officials used as part of their population management technique to showcase that the capital as capable of modernizing, at least on the surface.

Despite the introduction of new policies at the National School of Medicine that aimed to curb the traffic of corpses, problems existed inside state institutions that continued to challenge the

Porfirian vision of modernity. The situation unfolding at Hospital Juárez, a state institution, represented a tremendous threat to the vision state officials had for hospitals and public health. While state officials praised Hospital Juárez as the quintessential example of the Porfirian modernization, with the recent installation of electrical lighting, expansive wings, and the fact that it had hosted several important technological achievements in the medical field, such as the first application of radiography, problems remained.[75]

In particular, the idea that the hospital was a symbol of progress was laughable. According to the English language newspaper the *Mexican Herald*, Hospital Juárez was one of the most unsanitary and foul-smelling institutions in the capital. As the front page headline on June 10, 1903, read "Need for Reform at the Juárez Morgue; Dead Room is a Disgrace; Government Proposes to Eradicate Evils," wherein the anonymous reporter painted a vivid picture of how "modern" the hospital actually was. According to the article, the reporter's visit to Juárez left a lasting impression that "the city need not boast of keeping abreast of the times."[76] The reporter reserved particular disgust for how the hospital morgue stored corpses. Visitors, the author explained, never had to ask hospital employees for directions to the morgue because they could "smell it" as soon as they arrived on hospital grounds. The air was so pungent because, as the reporter wrote, the corpses in the morgue were "naked and foul-smelling, maimed with swollen stomachs and brown spots on their shoulders, threatening to burst as a result of decomposition." Furthermore, the author lamented the fact that the morgue was "a sickening sight and nauseating smell, which proves that no attention is given to this section of the hospital." While conditions in the morgue were poor, so too were those inside the hospital's renovated wings. The hospital boasted that expansive wings meant that doctors could see more patients. However, it failed to mention that it was a "truly defective and menacing environment for patients and employees," as patients afflicted with contagious diseases often shared hospital beds so that the hospital could continue to highlight that it treated a record number of patients.[77] The hospital had become the perfect microcosm for the realities of Porfirian modernization.

It remains unclear whether or not the Mexican government attempted to improve the conditions at Hospital Juárez after the *Mexican Herald*'s article. Nevertheless, the incident provides a unique window into the paradoxical reality that existed at state institutions during the Porfiriato. State officials sought to control the environment and employees at institutions like Hospital Juárez or the National School of Medicine, which was a hallmark of the population management technique promoted by Porfirian officials that emphasized progress at all costs, no matter the reaction of citizens. Yet the truth was far different. As the newspaper reporter discovered and others already knew, total control over all facets of citizens' lives, especially death and public health, was impossible. If a modern city was a developing organism, as many state officials believed, then it could neither be directed nor controlled in the chaos surrounding Porfirian modernization.

Conclusion

On the morning of June 20, 1916, cemetery worker Matias Aguilar was cleaning grave number 66 in lot C of Panteón Tepeyac, in a suburb several miles north of Mexico City, when he noticed something on the ground. Wedged between a headstone and grave curb was a glass jar wrapped in cloth. Attached to the jar was a note describing its contents: "I beg whoever finds this jar to carefully bury it, as it contains a legitimate and baptized child. I preserved the body in alcohol because I did not have enough resources to bury it. I left the jar in this holy place in the hands of those who will have mercy. God will pay charity."[78] After reading this, Aguilar realized he needed to let his cemetery administrator know what he had found so he could determine what to do next.

Upon hearing of the fetus, the cemetery administrator decided to contact legal medicine physician Roberto Caneda, who could examine the body. Caneda informed the administrator that to draw any conclusions, he first needed to perform an autopsy so that he could rule out the chance the infant was a victim of infanticide or contagious disease. Four days later, on June 24, Caneda submitted his report to the SSC, which outlined the fact that the jar contained a male infant measuring thirty-eight centimeters in length and at an intrauterine age of approximately seven months (twenty-eight weeks). Moreover, the organs, Caneda wrote, had been "in a fetal state and without notable alteration," which meant that the fetus had been healthy while inside the womb and had been on track for a normal birth. He concluded, however, that the mother delivered the infant prematurely (a typical pregnancy is forty weeks), thus resulting in his death seventy-two hours afterward.

While the rest of the case has disappeared from the historical record, it remains intriguing for several reasons. First, after Caneda had performed the autopsy, there appears to have been no investigation carried out by legal authorities to discover the child's parents. Second, the mother most likely had not given birth to the child at one of the city's hospitals since these institutions had specific protocols surrounding the disposal and burial of the dead, a glass jar not being one of them. Instead, state officials believed she had put herself and the city environment at risk by giving birth outside the clinical gaze of the hospital. Had the woman given birth at the hospital, officials assumed that the child would have survived his premature status with the help of modern medicine and technology. Third and most important, the case illustrated that traditional ways of mourning, including the burial site, remained integral to popular attitudes toward death and dying.[79] Panteón Tepeyac, where the cemetery worker found the fetus, was near La Basilica de Guadalupe, an important pilgrimage site in Mexican religious folklore where the Virgin Mary had appeared before peasant Juan Diego in 1555. While state officials tried repeatedly to secularize the process surrounding death in Mexico, moving burials, for example, out from under the umbrella of the Catholic Church and to the modern state, the reaction to death on the part of lower-class citizens often failed to match desired expectations. For those in the lower classes, death was becoming an increasingly expensive burden to families and relatives. Burial tickets, grave sites, coffins (rented or purchased), clothing, and accessories to adorn the grave site were all additional expenses that those living paycheck to paycheck could not afford. Thus, burials in state-sponsored, unmarked mass graves helped tether death for some to a final location that held significance for the families. Citizens continued to exert their autonomy in the face of an ever more stringent Mexican government that sought to pry into every facet of their lives.

Controlling how citizens dealt with death was a major component of how the Porfirian government believed it would achieve progress in Mexico City. The introduction of free burials for the city's poor population in public, secular cemeteries such as Panteón Dolores, officials believed, would help improve public health conditions and foster control over death. State officials considered the official approach to death, which included burial tickets, registration, death certificates and individual plots, an effective method for showcasing the government's benevolence and reverence for its citizens. However, the reality was that the disposal of dead bodies remained a contested space between citizen and state.

In particular, both the attempt to establish sanitary guidelines for disposing of the dead and the introduction of a bureaucratic process surrounding burials met resistance from lower-class citizens. For them, adopting these methods did not make sense. Rather than adopting these new systematic and detached methods of dealing with the dead, the urban poor preferred to deal with death in ways that demonstrated honor, respect, and reverence for the dead.

But for state officials, death was clinical, an inevitable event. However, among the lower classes, a moral economy surrounding death existed that state officials failed to take into consideration when it came to the introduction of rules and regulations related to death. The formation of a lower-class culture owed much to the help and support that individuals contributed. Thus, they were able to create a moral solidarity based on mutual assurance that everyone would look out for each other to ensure that traditions and customs survived.[80]

This tight-knit relationship represented a threat to the progress that Porfirian officials had tried to establish for more than a decade. Death was one arena that allowed these citizens an opportunity to circumvent official protocol. It was an event that permitted the marginalized to exert their autonomy in the face of an increasingly modernized capital that sought to erase them from the urban space, or at least to force them to adopt behavior considered civilized by the well-to-do. For all the attempts by state officials to control death and to create a cult of modern death in Mexico, death became one of the things that disobeyed them.[81] It was an event that neither science nor technology could subjugate, which resulted in the desire by state officials to continue to implement more stringent rules regarding death and burials as a way to protect public health. However, this attempt often failed, as regular citizens challenged how to process and interact with death by navigating (to their benefit) multiple spaces, such as cemeteries, hospitals, or even their own homes. The Porfirian state's continued reliance on its surveillance techniques and surveillance network, exemplified by frequent reports submitted to state officials by cemetery administrators and university officials, led to a growing defiance from the lower classes.[82] One of the areas that the lower classes could exercise some control over was how to interpret the rules surrounding dead bodies. This behavior, however, failed to act in accordance with the badge of modernity that President Díaz and his state officials had envisioned. Instead, it served as a reminder that the capital remained equal parts "barbarous" and modern.

Conclusion

From the perspective of foreign visitors, and much to the delight of Mexican state officials, the changes begun during the Díaz presidency to promote the capital as a home to the marvels of modernity had found an audience in the tourist guidebooks and literature that filled bookshelves worldwide in the early twentieth century. Foreign interest in the capital and in Mexico at large was instrumental for helping to improve Mexico's reputation at the time. These authors wined and dined with the elites of Mexico in an experience that state officials carefully constructed for them in order to present a sanitized image of Mexico, but the real Mexico was far different. One of the cities that state officials purposively chose to showcase was its capital, Mexico City. State officials had invested hundreds of millions of dollars into turning the city into a beacon of modernity that would rival London, Paris, Berlin, or Rome. Thus, President Porfirio Díaz and his government invited foreign writers to come visit Mexico, see it for themselves, and write favorable reports that could yield foreign economic opportunities through investment and tourism.

One such example were the writings of prolific British travel writer Ethel Tweedie, who visited the country several times during the Díaz presidency. In a 1906 publication, she informed her audience that "education, electric light, railways, splendid harbours, telegraphy, and endless modern inventions" were hustling aside the image of the old Mexico.[1] These changes, begun by Díaz, along with the government's commitment to science and technology, had supplanted the notion that all of Mexico remained a backward and barbarous nation. According to Tweedie, the fact that a significant number of indigenous citizens in Mexico City had given up excessive drinking along with their beliefs "in charms and weird cures for illnesses" was proof that Díaz and his state officials remained committed to improving the people of Mexico. Moreover, advancements in public health and sanitation was transforming the capital into a city with "finest combination of ancient and modern architecture in the world, clean streets, well-paved, well-lighted roads, good police force, and excellent tram service," which was instrumental to putting it on par with the more revered European capitals.[2] This reputation would not have been possible without the help of President Díaz and his state officials, who dedicated themselves to the creation of a regulated city environment guided by science and technology and complemented by the introduction of population management techniques that sought to control how citizens interacted with death on a daily basis.

Yet in the rush to crown Mexico City as the Paris or London of the Western Hemisphere, visitors and state officials chose to ignore the realities of daily life in the capital. The images and improvements to the landscape as described by Tweedie and others were in fact sanitized to satisfy the desires of state officials, elite residents, and foreign tourists who sought a particularly romantic version of Mexico and its cities. While the capital could boast of electric lights, wide boulevards, electric trams, and open green spaces for walking, if visitors or residents veered slightly off a prescribed course laid out in guidebooks, they would run into the true version of the capital. For example, numerous decomposing corpses found outside corpse deposits, inside and outside of cemetery walls, on board electric trams, and at visitors' feet were within a stone's throw from popular routes. Additionally, all visitors and residents had to do was breathe in the surrounding air, especially near hospitals, to inhale the stench of death that remained pungent in the capital.[3] The real Mexico City was far from the modern image constructed by state officials, elite residents, and travel writers. The city suffered from a multitude of hygienic problems, least of which was the lack of opportunities lower-class citizens had to bathe, despite the supposed success that state officials championed in the name of public health improvements.

Studies of public health and modernization during the presidency of Porfirio Díaz have explained how the government sought to extend its control into the arena of citizens' health.[4] By focusing on death, my work has revealed how pervasive the corpse problem was in the capital and how the dead threatened the image of progress and health that state officials had constructed during the late nineteenth and early twentieth century. Furthermore, the corpse problem also illustrates the extent to which lower-, middle-, and upper-class citizens constantly negotiated how best to handle the rules surrounding death in a modern society. State officials' attempts to improve public health and define a modern death were integral parts of the population management technique begun during the presidency of Porfirio Díaz. According to state officials, how citizens buried, mourned, and interacted with the dead was essential to the campaign to improve health and hygiene. Dying in the Porfirian era was no longer a private matter. The living, especially state officials and well-to-do citizens, sought to play a fundamental role in the construction of modern citizenship. Science and technology bound the state, citizen, and dead in a Porfirian milieu that focused on showcasing how modern the capital had become under Díaz's leadership.

Nevertheless, tensions emerged during this citizenship project as lower-class citizens often chose to resist rather than conform to expected behaviors. Official state transportation methods for the dead, such as the railroad, corpse carriage, and electric tram, were all symbols of modernity that meant different things to various citizens. For example, state officials and the well-to-do saw these technological changes as integral to changing the culture of Mexico City. But for the lower classes, these new technologies meant little to their daily lives and failed to change the way they chose to interact with death. Nevertheless, President Díaz made sure to introduce new rules and regulations in 1887 that would govern the hygienic transportation of corpses in an effort to protect public health, a key feature that Porfirian officials argued had not existed before. To them, the ability to link new technology to population management techniques and public health was vital for demonstrating that the capital was capable of becoming modern. The Porfirian vision for how residents would use this technology, however, encountered problems early on. Corpse deposits located throughout the capital and connected to railways or electric tram lines, for example, filled beyond capacity faster than officials had anticipated. Once the deposits filled, lower-class residents left corpses in the streets nearby. Rather than take the corpses back home, which meant paying for a death certificate, burial ticket, and burial transportation costs, these individuals and families chose to do what was convenient for them. The result, however, was a rapid explosion of unsanitary environments that exacerbated public health conditions and threatened the image of progress desired by state officials.

In addition to new transportation methods introduced by the Mexican government, state officials also turned their attention to revising medical education as part of the Porfirian discourse on modernity. The government, medical institutions, and medical professionals all worked together to improve public health in Mexico City in order to make healthier citizens, which would help change the perception of the capital, and maybe the country, from backward to modern. This approach to population management had an influential ally in Dr. Eduardo Licéaga, President Díaz's personal physician, friend, head of the Superior Sanitation Council, and director of the National School of Medicine. He was an instrumental figure in the campaign to improve medical education in the 1890s since he believed doctors needed better training in order to save the lives of patients and to offer sage advice about health. There were too many deaths during the Porfirian era that Licéaga believed would have been preventable if medical training received a boost into the modern pedagogical world. Mexican medicine lagged behind other major countries, in Licéaga's opinion, primarily because the curricula focused too much on medical theories and rote learning from textbooks. Dissection, Licéaga believed, was the key to solving Mexico's health problems. With his social standing and influence in Porfirian society, Licéaga was able to use his alma mater, the National School of Medicine, as the laboratory for experimenting with changes to the curriculum before applying them to all medical institutions.

Thus, from 1893 to 1910, Licéaga used his influence to urge faculty members to abandon the methods of instruction they had received as students and look instead to the human body as inspiration. Dissection provided the foundation students needed, and cadavers were their best textbooks. This change to medical education received substantial support from state officials, medical physicians, professors, students, and, of course, President Díaz himself. The resources committed to medical professionals and unwavering support from high-ranking officials reinforced the fact that state officials wanted to use improved medical education as part of the process to regulate the bodies of citizens. Embracing medicine and the sage advice of medical professionals was another step that state officials believed was needed not only to improve individual health but to improve the health of the capital and, officials hoped, the nation.

But the attempt to turn the capital into a modern landscape free of tremendous public health problems required more than just new transportation methods for the dead or a revamped medical education system. The desire by state officials to protect public health, which included removing the scattered corpses found near corpse deposits, led to the introduction of a new industry in Mexico: funerary technology. New technologies designed specifically for the treatment of dead bodies emerged in the first decade of the twentieth century at a rate that far surpassed that of the late nineteenth century. The appearance of such technology development coincided with the rest of the changes underway in the capital that sought to create a safe, healthy, and organized city, free from decomposing corpses and the threat they posed to public health. This new funerary technology led to the development of five distinct methods of body disposition: coffins, burial vaults, topical embalming, arterial embalming, and cremation. All of these methods exacerbated the existing class division in the city, especially since the audience for most of the new methods was elite and middle-class citizens. For lower-class citizens, the only option for their bodies was cremation, a method chosen for them by state officials and elite citizens who wanted a more effective and hygienic method for removing the growing number of lower-class corpses found in city streets, corpse deposits, hospitals, and other state institutions. These corpses, usually in various stages of putrefaction, threatened to ruin the modern image of Mexico City that President Díaz and his científicos had worked hard to create; thus, the state required total elimination, vaporizing their bodies into dust so that no trace remained.

Funerary technology was an important ingredient in the Porfirian population management recipe. New and innovative methods for disposing of bodies were integral to helping state officials create a definition of modern death based on their interpretation of what constituted a safe and hygienic approach. For state officials, practices surrounding death and the dead were important for inculcating behaviors and customs that they considered modern and no longer backward or uncivilized. But rather than exercising patience with the behaviors surrounding death that state officials expected lower-class citizens to adopt, officials grew frustrated by the glacial pace of how the changes unfolded. The speed of change had little to do with lower-class citizens' inability to change or failure to understand. Instead, the urban poor chose to ignore the changes, ignore the rules, and continue living their lives in ways that made sense to them. Whether, for example, it was by burying a loved one illegally on the grounds of a church cemetery rather than a public, secular cemetery or by using the city's tram system and corpse collection to their advantage by dropping off dead bodies without the proper paperwork, these citizens made a choice to exercise their independence in the face of a growing and invasive state government.

By doing so, they undermined the state's authority. Continuing to rely on traditional body disposition methods and unhygienic practices and maintaining an intimacy with death and dying that officials considered unhealthy demonstrated how lower-class citizens defined their identities in the face of an increasingly powerful state apparatus that sought to control all facets of their lives. By focusing on the dead, this study has emphasized that in personal matters such as sex or death, citizens did not want the government to define them according to rules they did not control or construct themselves. These individuals continued to react to various situations based on what was best for them, even if it meant that state officials and elite citizens viewed them with disdain and as living relics of the old forgotten Mexico.

Hustling the old Mexico aside, with its people, beliefs, traditions, and spirit, as state officials had hoped to accomplish, proved too difficult a task to achieve. Díaz and his state officials believed that nature and culture were distinct categories that they could manipulate in their favor, namely to push for modernity and all of its attributes, especially better sanitation and hygiene in order to create more perfect public health in Mexico City. For them, this approach to modernization was central to establishing an official narrative for the capital to eliminate the chaotic past of the nineteenth century and replace it with one that emphasized the importance of science, medicine, and technology, all hallmarks of Porfirian Mexico. But Porfirio Díaz and his state officials encountered resistance at every turn while trying to get citizens to accept the state's modernization efforts. The past, as sociologist Bruno Latour has argued, could not disappear entirely to make room for a present that used rationality, scientific truth, and technology.[5] The only citizens who chose to embrace the changes that would affect their daily lives were elite and middle-class citizens. For the urban poor, however, the focus on improving public health through new transportation methods for the dead, an emphasis on dissection in medical schools, or new funerary technology, failed to impact their lives in any notable way. Thus, the attempts to capture and define what it meant to be modern in Mexico City were examples of the metaphoric dog that chases its own tail: an action that is never able to deliver the desired result. Despite the efforts of state officials and the Mexican government to modify the landscape of the capital as well as the behavior of its citizens by introducing various population management techniques, the result was the creation of a capital that remained divided and incapable of becoming modern.

NOTES

Introduction

1. Jason Beaubien, "In Diabetes Fight, Lifestyle Changes Prove Hard to Come by in Mexico," NPR, April 7, 2017, http://www.npr.org/sections/goatsandsoda/2017/04/07/522596080/in-diabetes-fight-lifestyle-changes-prove-hard-to-come-by-in-mexico.

2. Alegre-Díaz, et al., "Diabetes and Cause-Specific Mortality in Mexico City," 1961–71; Lerman-Garber, Gómez-Pérez, and Quibrera-Infante, "Mexico," 195–204.

3. Beaubien, "In Diabetes Fight."

4. Jason Beaubien, "How Diabetes Got to Be the No. 1 Killer in Mexico," NPR, April 8, 2017, http://www.npr.org/sections/goatsandsoda/2017/04/08/522184483/pork-tacos-topped-with-fries-fuel-for-mexicos-diabetes-epidemic.

5. Jason Beaubien, "Pork Tacos Topped with Fries," NPR, April 5, 2017, http://www.npr.org/sections/goatsandsoda/2017/04/05/522038318/how-diabetes-got-to-be-the-no-1-killer-in-mexico.

6. Agostoni, *Monuments of Progress*, 38.

7. Tweedie, *Porfirio Díaz*, 265.

8. Scott, *Seeing Like a State*, 92.

9. Sloan, *Death in the City*; Ruiz-Alfaro, "Threat to the Nation," 41–62; Bliss, *Compromised Positions*; and Rubenstein, *Bad Language, Naked Ladies*.

10. Archivo General de la Nación (hereafter AGN), Fondo-Tribunal Superior de Justicia del Distrito Federal (hereafter F-TSJDF), caja 1409, expediente 248896, November 5, 1916, 13–14.

11. AGN, TSJDF, caja 1409, expediente 248896, November 5, 1916, 13–14.

12. AGN, TSJDF, caja 1409, expediente 248896, November 5, 1916, 18–19.

13. Scott, *Moral Economy*, 43–44.

14. Sowell, "Quacks and Doctors," 15. For more, see Leiby, "Royal Indian Hospital," 573–80; and Hernández and Foster, "Curers and Their Cures," 19–46.

15. Lomnitz, *Death and the Idea*, 232–33.

16. Lomnitz, 241; Knight, *Mexican Revolution*, 2:42–44. For more on cofradías in Mexico City, see Larkin, "Confraternities and Community," 189–214; García-Ayluardo, "Confraternity, Cult, and Crown."

17. Knight, *Mexican Revolution*, 2:43.

18. Lomnitz, *Death and the Idea*, 261.

19. Brandes, *Skulls to the Living*, 19–42; and Viquiera-Alban, "El sentimiento," 27–63.

20. Voekel, *Alone before God*, 72.

21. Voekel, 73–74.

22. Voekel, 73–74.

23. Voekel, 106–22.

24. McCrea, *Diseased Relations*, 70–75.

25. Hale, *Mexican Liberalism*, 125.

26. Kirkwood, *History of Mexico*, 103–4. A similar struggle over control of death occurred in Vienna, the capital of the Hapsburg Empire. In 1867 the first professional funeral company replaced the traditional funerary services of Catholic church sextons, which, like its name, Enterprise des Pompes Funèbres, promised to provide elegant burial services that matched the tastes of the well-to-do. For more, see Buklijas, "Culture of Death," 570–607.

27. Lewis, *Tepoztlán*, 83–85; Brandes, *Skulls to the Living*.

28. Tweedie, *Porfirio Díaz*, 2.

29. Castronovo, *Death, Eroticism*, xii.

30. Verdery, *Political Lives*, 52.

31. Verdery, 29.

32. Rosen, *History of Public Health*, lxxix.

33. McKeown had worked on this argument over a series of decades in articles that focused on the issues he believed had contributed to improved public health, culminating in the publication of *The Modern Rise of Population*. The articles that helped to shape his book were: McKeown and Brown, "Medical Evidence"; McKeown and Record, "Reasons for the Decline"; McKeown, Brown, and Record, "An Interpretation"; and McKeown, Record, and Turner, "An Interpretation."

34. Foucault, *Birth of the Clinic*, xii. Also see Foucault, *Madness and Civilization*; Foucault, *Discipline and Punish*; and Foucault, *History of Sexuality*.

35. Scott, *Weapons of the Weak*, xv–xvi.

36. Scott, *Seeing Like a State*, 82.

37. Latour, *We Have Never Been*. Latour's work borrows from Raymond Williams's essay on "Ideas of Nature," wherein Williams argues that nature is not a singular unchanging entity that can be separated from humans. For more, see Williams, "Ideas of Nature," in Williams, *Culture and Materialism*, 67–85.

38. Piccato, *City of Suspects*. Criminalization of the poor also occurred during the seventeenth and eighteenth centuries when Mexico City leaders attempted to sweep beggars and vagrants off the streets and classify them according to their worthiness to the state. For more, see Arrom, *Containing the Poor*.

39. Bliss, *Compromised Positions*.

40. Lomnitz, *Death and the Idea*.

41. Esposito, *Funerals, Festivals*.

42. López, "Cadaverous City."

43. Voekel, "Peeing on the Palace," 183–208; Staples, "Policia y Buen Gobierno," 115–26.

44. Carrillo, "Médicos del México decimonónico," 351–75.

45. Carrillo, "Profesiones sanitarias."

46. Agostoni, *Monuments of Progress*.

47. Agostoni, "'Que no traigan,'" 97–120.

48. López, "Cadaverous City," 10.

49. Voekel, *Alone before God*; Kapelusz-Poppi, "Rural Health," 261–83; Cassity, "Health, Sanitation, Hygiene"; and McCrea, *Diseased Relations*.

50. Kemper and Peterson Royce, "Mexico Urbanization since 1821," 267–89; Kemper and Peterson Royce, "Urbanization in Mexico," 93–128.

51. Knight, *Mexican Revolution*, 1:15.

52. Castronovo, *Death, Eroticism*, xii.

53. Castronovo, 4.

54. Verdery, *Political Lives*, 52.

55. Verdery, 29.

56. Forman, "On the Historical Forms," 60, 72, 83. For an excellent example discussing the increasing use of the term professionalization among physicians in the United States, see Sappol, *Traffic of Dead Bodies*.

57. McKiernan-González, *Fevered Measures*, 11.

58. Shah, *Contagious Divides*, 5.

59. Spivak, "Can the Subaltern Speak?" 271–316.

60. For more on the relationship that existed between the government, public health, and dead bodies in Mexico, China, Japan, the United States, and Sweden, see Voekel, *Alone before God*; Rogaski, *Hygienic Modernity*; Bernstein, *Modern Passings*; Gilpin Faust, *This Republic of Suffering*; and Åhrén, *Death, Modernity*.

61. González Navarro, *Estadisticas Sociales del Porfiriato*, 9.

62. McCrea, *Diseased Relations*, 4.

1. Moving into the Modern Era

1. Tweedie, *Porfirio Díaz*, 317–21.

2. Agostoni, *Monuments of Progress*, 47.

3. Matthews, *Civilizing Machine*, 23–24.

4. McCaa, "Peopling of 19th Century," 606.

5. Van Hoy, *A Social History*, 1.

6. Blum, *Domestic Economies*, 12.

7. Blum, 12–14. For more, see González Navarro, *Estadisticas Sociales del Porfiriato*.

8. Lear, "Mexico City," 444–92.

9. For more, see Andrews, *Afro-Argentines* and Andrews, *Blacks and Whites*. In late nineteenth- and early twentieth-century Latin American port cities, such as Buenos Aires or São Paulo, governments instituted new immigration laws designed to attract white immigrants from eastern Europe in order to spark racial mixing and eliminate the presence of African and indigenous blood from future generations.

10. Matthews, *Civilizing Machine*, 7–8.

11. Van Hoy, *Social History*, 167. All Mexican peso to U.S. dollar conversions contained within this book are based on data available in *Estadísticas Históricas de México, vol. 2: cuadro 21.6*. To determine the worth of this amount in 2018 dollars, I used https://www.measuringworth.com/calculators/uscompare/, which allows for calculations based on historical monetary values. All calculations made are rough estimates only. For more information on conversions, read the essay by Lawrence H. Officer and Samuel H. Williamson on their website titled "Explaining the Measures of Worth" (2018). The website contains a wealth of information for anyone looking to get the most reliable historical economic data as well as an emphasis on nominal measures of monetary values. For peso values, see Denzel, *Handbook of World Exchange Rates*, 491.

12. Piccato, *City of Suspects*, 23–24.

13. Blum, *Domestic Economies*, 6.

14. Agostoni, *Monuments of Progress*, 26.

15. McCaa, "Peopling of 19th Century," 6–7. Life expectancy in other large urban cities like London (between 44 and 49 years) and Paris (between 46 and 50 years) nearly doubled that of Mexico City. For more, see Floud and Johnson, *Cambridge Economic History*.

16. Piccato, *City of Suspects*, 26–28.

17. Janvier, *Mexican Guide*, 136.

18. Agostoni, *Monuments of Progress*, 25–30.

19. Piccato, *City of Suspects*, 13–33. This was not the first time the government had attempted to alter the behavior and customs of the lower classes. During the eighteenth century, Bourbon reformers attempted to instill behavior they considered "proper" in the urban poor through a series of hygienic measures that government officials believed would lead to improvement for their health and the health of the city. For more, see Voekel, "Peeing on the Palace," 183–208; and Staples, "Policia y Buen Gobierno," 115–26.

20. Piccato, *City of Suspects*, 19–21; and Blum, *Domestic Economies*, 14–15.

21. Cohen, *A Body Worth Defending*, 21–22.

22. Voekel, *Alone before God*, 92.

23. Voekel, 106–45; and McCrea, *Diseased Relations*, 70–75.

24. Cohen, *A Body Worth Defending*, 116–29.

25. Matthews, *Civilizing Machine*, 48.

26. Piccato, *City of Suspects*, 13; and Mraz, *Looking for Mexico*, 31.

27. Matthews, *Civilizing Machine*, 4.

28. Matthews, 50.

29. Van Hoy, *Social History*, 16.

30. Archivo Histórico de la Secretaría de Salubridad y Asistencia (hereafter AHSSA), Fondo-Salubridad Pública (hereafter F-SP), Sección-Medicina Legal (hereafter Sec-ML), caja 2, expediente 24, November 6, 1877, 1–7.

31. AHSSA, F-SP, Sección-Secretaría de Justicia (hereafter Sec-SJ), caja 4, expediente 5, March 11, 1887, 1.

32. AHSSA, F-SP, Sec-SJ, caja 4, expediente 5, March 11, 1887, 1–2.

33. AHSSA, F-SP, Sec-SJ, caja 4, expediente 5, March 11, 1887, 1.

34. Burns, Acuna-Soto, and Stahle, "Drought and Epidemic Typhus," 1.

35. Nelson, "Yellow Fever," 397.

36. AHSSA, F-SP, Sec-SJ, caja 4, expediente 5, March 11, 1887, 1–2.

37. Cohen, *A Body Worth Defending*, 97.

38. Matthews, "De Viaje," 267.

39. Rivera Cambas, *Mexico pintoresco*, 69–70.

40. Rivera Cambas, 70.

41. Rivera Cambas, 70.

42. Alcaraz Hernández, "Las pestilentes," 95.

43. Alcaraz Hernández, 95.

44. Alcaraz Hernández, 95–96.

45. Alcaraz Hernández, 96.

46. Archivo Histórico de la Ciudad de México (hereafter AHCM), Fondo-Ayuntamiento de México/Gobierno del Distrito Federal (hereafter F-AM/GDF), Serie-Hospital de San Pablo (hereafter S-HSP), caja 1, expediente 19, March 26, 1889, 1.

47. AHCM, F-AM/GDF, S-HSP, caja 1, expediente 19, March 29, 1889, 3.

48. AHCM, F-AM/GDF, S-HSP, caja 1, expediente 19, March 29, 1889, 2–3.

49. Cohen, *A Body Worth Defending*, 116.

50. Lear, *Workers, Neighbors, and Citizens*, 33–36.

51. Van Hoy, *Social History* xviii–xxii; and Piccato, *City of Suspects*, 1–3.

52. Coatsworth, *Growth against Development*, 6–8. Also see Rohlfes, "Police and Penal Correction," 1–4.

53. Tenorio-Trillo, *Mexico at the World's*, 25; Buffington and French, "Culture of Modernity," 404–7.

54. AHCM, F-AM/GDF, Serie-Panteones (hereafter S-P), caja 1, expediente 19, April 26, 1894, 27–28. Conversion based on real price commodity value using exchange rate for the year 1894 of 1.979 pesos to $1.

55. Alcaraz Hernández, "Las pestilentes," 97.

56. Alcaraz Hernández, "Las pestilentes," 97–98.

57. AHCM, F-AM/GDF, S-HSP, caja 1, expediente 19, July 18, 1889, 4.

58. Matthews, "De Viaje," 274; and French, *Peaceful and Working People*, 110–11.

59. Cohen, *A Body Worth Defending*, 172.

60. Beezley, *Mexico in World History*, 96. For more see Beezley, *Judas at the Jockey*.

61. Garza, *Imagined Underworld*, 12–36.

62. AHCM, F-AM/GDF, S-HSP, caja 1, expediente 19, July 18, 1889, 4. His complaint included residents of the following neighborhoods: Plaza de Jardín, Villamil, Carbonero, and Santa María la Redonda.

63. AHCM, F-AM/GDF, S-P, caja 1, expediente 19, July 25, 1889, 10.

64. Prantl and Grosó, *La Ciudad de México*, 922.

65. AHCM, F-AM/GDF, S-P, caja 1, expediente 19, July 25, 1889, 10.

66. Hale, *Transformation of Liberalism*, 206–42.

67. Piccato, *City of Suspects*, 50–63.

68. Rohlfes, "Police and Penal Correction," 209–11; Piccato, *City of Suspects*, 61–62.

69. Terry, *Terry's Mexico*, 369.

70. Buffington, *Criminal and Citizen*, 13–14; and Piccato, *City of Suspects*, 14. For similar views of prisoners and lower-class citizens in the United States, England, and Australia, see Sappol, *Traffic of Bodies*; Richardson, *Death, Dissection*; and MacDonald, *Human Remains*.

71. AHCM, F-AM/GDF, S-P, caja 1, expediente 19, August 1, 1889, 14.

72. Janvier, *Mexican Guide*, 78–79.

73. AHCM, F-AM/GDF, S-P, caja 1, expediente 19, August 1, 1889, 15.

74. AHCM, F-AM/GDF, S-P, caja 1, expediente 19, August 1, 1889, 15.

75. AHCM, F-AM/GDF, S-P, caja 1, expediente 19, January 29, 1890, 17.

76. AHCM, F-AM/GDF, S-P, caja 1, expediente 19, February 12, 1894, 22–23.

77. Tenorio-Trillo, *Mexico at the World's*, 30–31.

78. Cohen, *A Body Worth Defending*, 103–4.

79. Tenenbaum, "Streetwise History," 139–40. Tenenbaum argues that placing monuments in prominent locations, such as exclusive neighborhoods near Paseo de la Reforma, reinforced the elites' belief in their own superiority and role in guiding Mexico's future. Likewise, state officials purposefully chose the location of the railway and corpse deposits in popular neighborhoods to keep the symbols of progress on constant display for citizens.

80. For an additional example of the role of Porfirian pedagogy and modernization in Mexico City, see Esposito, "Death and Disorder," 106–19.

81. AHCM, F-AM/GDF, S-P, caja 3460, expediente 871, June 14, 1898, 1.

82. AHCM, F-AM/GDF, S-P, caja 3460, expediente 871, June 14, 1898, 3.

83. AHCM, F-AM/GDF, S-P, caja 9, expediente 855, June 26, 1903, 1. For more on the role of the Superior Sanitation Council, see Agostoni, *Monuments of Progress*, 57–76; Ross, "From Sanitary Police."

84. AHCM, F-AM/GDF, S-P, caja 9, expediente 855, June 26, 1903, 1.

85. AHCM, F-AM/GDF, S-P, caja 9, expediente 855, June 26, 1903, 1.

86. AHCM, F-AM/GDF, S-P, caja 9, expediente 855, June 26, 1903, 1.

87. Lomelí Vanegas, "Ciencia Económico y Positivismo," 199–222.

88. AHCM, F-AM/GDF, S-P, caja 9, expediente 855, July 8, 1903, 2.

89. French, "Progreso Forzado," 191–207.

90. Cohen, *A Body Worth Defending*, 176.

91. Pitt, *Walks through Lost Paris*, 5.

92. Pitt, 5–8.

93. Scott, *Seeing Like a State*, 63, 81–83.

94. AHCM, F-AM/GDF, S-P, caja 15, expediente 1408, May 10, 1905, 1.

95. AHCM, F-AM/GDF, S-P, caja 15, expediente 1408, May 12, 1905, 3.

96. Pilcher, *Sausage Rebellion*, 41–87.

97. Pilcher, 60; Piccato, *City of Suspects*, 28–29.

98. AHCM, F-AM/GDF, S-P, caja 15, expediente 1408, May 21, 1905, 7.

99. AHCM, F-AM/GDF, S-P, caja 15, expediente 1408, July 29, 1905, 3–4; and AHCM, F-AM/GDF, S-P, caja 1, expediente 19, August 1, 1889, 15.

100. Reese and Reese, "Revolutionary Urban Legacies," 361–73.

101. Tenorio-Trillo, *Mexico at the World's*, 20.

102. Miles, *World's Fairs*, 80.

103. Tenorio-Trillo, *Mexico at the World's*, 64.

104. This is the historical opportunity cost for 1889 based on 1.334 pesos to the dollar conversion rate.

105. Tenorio-Trillo, *Mexico at the World's*, 65.

106. Tenorio-Trillo, 75.

107. Tenorio-Trillo, 73. For more on Lord Kingsborough, see Whitmore, "Lord Kingsborough," 8–16.

108. Tenorio-Trillo, *Mexico at the World's*, 75.

109. AHCM, F-AM/GDF, Serie-Hospital San Pablo (hereafter S-HSP), caja 17, expediente 1526, July 29, 1905, 3–4. For more on traditional Mexican architecture, see Sturgis, *Dictionary of Architecture*, 895–907; Kalach, "Architecture and Place," 114.

110. For more on rabies in Mexico, see Ramírez de Arrellano, "Higiene: Profilaxis de la rabia," 206–9. For more on the problem between feral dogs, bodies, and city residents in Mexico City, see Tenorio-Trillo, *I Speak*.

111. AHCM, F-AM/GDF, S-P, caja 17, expediente 1526, July 29, 1905, 6. For more on the common hygienic features of architecture in the era, see Van Der Bent, *Problem of Hygiene*, 295; and Barnard, *School Architecture*, 40–62.

112. AHCM, F-AM/GDF, S-P, caja 17, expediente 1526, July 29, 1905, 3–4.

113. Taylor, *Principles of Scientific Management*.

114. Cynthia Crossen, "Early Industry Expert Soon Realized a Staff Has Its Own Efficiency," *Wall Street Journal*, November 6, 2006, B1.

115. Thompson, "Time, Work-Discipline," 56–97.

116. Daft, *Organization Theory and Design*, 23.

117. Haber, *Efficiency and Uplift*. For more on Taylor and the issue of modernizing workers, see Smith, "Political Economy of Pacing"; Nelson, *Frederick W. Taylor*; Sabel, *Work and Politics*; Sullivan, "Challenge of Economic Transformation"; and Kanigel, *One Best Way*.

118. Taylor, *Principles of Scientific Management*, 29.

119. This is the historic real price based on 2.018 pesos to the dollar for 1905.

120. "Funeral Cars in Mexico." This is the historic real price based on 2.169 pesos to the dollar for 1898.

121. Agostoni, *Monuments of Progress*, 85–87. Conversion for Noriega's carriages is based on real price using the year 1905. For the public works projects, this is the historical opportunity cost based on the average peso value between the years 1877 and 1910 of 1.64 pesos to the dollar using the year 1893 for the conversion.

122. Kirkwood, *History of Mexico*, 130.

123. Agostoni, *Monuments of Progress*, 87. For the public works projects, the conversion is based on historical opportunity cost using the year 1893.

124. Toledo Martínez, "Historia social," 29–31. On Montevideo, see Rosenthal, "Arrival of the Electric"; on Buenos Aires, see García Héras, *Transportes, negocios y política*; and on Rio de Janeiro, see Boone, "Streetcars and Politics."

125. Cooke, "An Extreme Power Engineer." This is the historical opportunity cost based on the average peso value for the years 1898 to 1910 of 2.0875 pesos to the dollar and the year 1898 as the initial year for monetary conversion.

126. Pilcher, *Sausage Rebellion*, 72–73.

127. "Frederick Pearson Starks"; Toledo Martínez, "Historia social," 98–99.

128. Toledo Martínez, "Historia social," 45.

129. Piccato, *City of Suspects*, 24–25.

130. Toledo Martínez, "Historia social," 99.

131. Matthews, *Civilizing Machine*, 145–46.

132. Lomnitz, *Death and the Idea*, 377–81.

133. AHCM, F-AM/GDF, S-P, caja 29, expediente 2591, March 13, 1909, 1.

134. Agostoni, *Monuments of Progress*, 83.

135. Martin, *Mexico of the Twentieth*, 160–77.

136. Tenorio-Trillo, "1910 Mexico City," 90.

137. AHCM, F-AM/GDF, S-P, caja 29, expediente 2591, March 13, 1909, 1.

138. López, "Cadaverous City," 50.

139. AHCM, F-AM/GDF, S-P, caja 29, expediente 2591, March 13, 1909, 2.

140. López, "Cadaverous City," 55–56.

141. López, 55–56. This conversion is based on real price commodity using 1.282 pesos to the dollar for the year 1887.

142. "Mining States of Mexico," *Overland Monthly* 56 (July–December 1910), 64.

143. de la Torre Rendón, "Las imágenes fotográficas," 357.

144. AHCM, F-AM/GDF, S-P, caja 29, expediente 2591, March 13, 1909, 1.

145. AHCM, F-AM/GDF, S-P, caja 29, expediente 2591, March 19, 1909, 1.

146. AHCM, F-AM/GDF, S-P, caja 29, expediente 2591, March 19, 1909, 1–2.

147. AHCM, F-AM/GDF, S-P, caja 29, expediente 2591, July 7, 1909, 7.

148. AHCM, F-AM/GDF, S-P, caja 29, expediente 2591, July 7, 1909, 8.

149. AHCM, F-AM/GDF, S-P, caja 29, expediente 2591, July 13, 1909, 9.

150. AHCM, F-AM/GDF, S-P, caja 32, expediente 2817, July 29, 1910, 1.

151. AHCM, F-AM/GDF, S-P, caja 32, expediente 2817, July 29, 1910, 1–2.

152. Terry, *Terry's Mexico*, 256.

153. AHCM, F-AM/GDF, S-P, caja 32, expediente 2817, August 3, 1910, 2.

154. AHCM, F-AM/GDF, S-P, caja 32, expediente 2817, August 12, 1910, 7.

155. AHCM, F-AM/GDF, S-P, caja 32, expediente 2817, August 12, 1910, 8.

156. Terry, *Terry's Mexico*, 407; AHCM, F-AM/GDF, S-P, caja 32, expediente 2843, August 4, 1911, 1.

157. AHCM, F-AM/GDF, S-P, caja 32, expediente 2843, August 4, 1911, 1–2.

158. AHCM, F-AM/GDF, S-P, caja 32, expediente 2843, August 4, 1911, 2.

159. AHCM, F-AM/GDF, S-P, caja 32, expediente 2878, December 24, 1912, 2.

160. AHCM, F-AM/GDF, S-P, caja 3472, expediente 260, January 28, 1918, 29. For more on the dangers of early automobiles in Mexico City, see Piccato, *City of Suspects*, 100–101.

161. This conversion is based on historical labor earnings for 1918, when 1.807 pesos equaled a dollar.

162. AHCM, F-AM/GDF, S-P, caja 3472, expediente 260, January 28, 1918, 24; AHCM, F-AM/GDF, S-P, caja 3472, expediente 260, July 20, 1918, 18; AHCM, F-AM/GDF, S-P, caja 3472, expediente 260, October 16, 1918, 8; AHCM, F-AM/GDF, S-P, caja 3472, expediente 260, December 2, 1919, 33.

163. Nas and De Giosa, "Conclusion."

2. "An Extraordinary Tool"

1. Birn and Carrillo, "Neighbours on Notice."

2. Baker, *Sanitation in 1890*, 4.

3. Ross, "Mexico's Superior Health Council," quoted on page 573.

4. Agostoni, *Monuments of Progress*, 59.

5. Ross, "Mexico's Superior Health Council," 579. For a larger discussion on clientelism, see Robinson and Verdier, "Political Economy of Clientelism," 260–91.

6. Johns, *City of Mexico*, 33–34.

7. Archivo Histórico de la Universidad Nacional Autónoma de México (hereafter AHUNAM), Fondo-Escuela Nacional de Medicina (hereafter F-ENM), Ramo-Dirección (hereafter R-D), Subramo-Secretaría (hereafter Sub-Sec) Serie-Correspondencia General (hereafter S-CG), caja 25, expediente 113, 1912, 1127–39.

8. AHUNAM, F-ENM, R-D, Sub-Sec, S-PG, caja 31, expediente 21, March 30, 1889, 100–101.

9. AHUNAM, F-ENM, R-D, Sub-Sec, S-PG, caja 31, expediente 21, March 30, 1889, 102–3.

10. AHUNAM, F-ENM, R-D, Sub-Sec, S-PG, caja 31, expediente 21, March 30, 1889, 100–103.

11. AHUNAM, F-ENM, R-D, Sub-Sec, S-PG, caja 31, expediente 21, March 30, 1889, 103–6.

12. AHUNAM, F-ENM, R-D, Sub-Sec, Serie-Programa de Estudios (hereafter S-PE), caja 18, expediente 7, June 20, 1893, 41; Potter, "American Public Health Association," 114.

13. Buklijas, "Culture of Death," 472.

14. AHUNAM, F-ENM, R-D, Sub-Sec, S-PE, caja 18, expediente 7, June 20, 1893, 42.

15. AHUNAM, F-ENM, R-D, Sub-Sec, S-PE, caja 18, expediente 7, June 20, 1893, 42.

16. AHUNAM, F-ENM, R-D, Sub-Sec, S-PE, caja 18, expediente 7, June 20, 1893, 43.

17. AHUNAM, F-ENM, R-D, Sub-Sec, S-PE, caja 18, expediente 7, June 20, 1893, 62.

18. For more on medical instruments, see Cunningham and Williams, *Laboratory Revolution in Medicine*; and Starr, *Social Transformation*.

19. Agostoni, *Monuments of Progress*, 86–88.

20. AHUNAM, F-ENM, R-D, Sub-Sec, S-PE, caja 18, expediente 7, June 20, 1893, 63.

21. AHUNAM, F-ENM, R-D, Sub-Sec, S-PE, caja 18, expediente 7, June 20, 1893, 63.

22. AHUNAM, F-ENM, R-D, Sub-Sec, S-PE, caja 18, expediente 7, June 20, 1893, 65.

23. AHUNAM, F-ENM, R-D, Sub-Sec, S-PE, caja 18, expediente 7, June 20, 1893, 65.

24. AHUNAM, F-ENM, R-D, Sub-Sec, S-PE, caja 18, expediente 7, June 20, 1893, 68–71.

25. AHUNAM, F-ENM, Ramo-Institutos y Sociedades Medicas (hereafter R-ISM), Serie-Museo Anatomia-Patológica (hereafter S-MAP), caja 40, expediente 1, January 14, 1895, 1.

26. Sappol, "'Morbid Curiosity.'"

27. Knox, *Anatomist's Instructor*, 7.

28. AHUNAM, F-ENM, R-ISM, S-MAP, caja 40, expediente 1, January 14, 1895, 1.

29. Sappol, *Traffic of Dead Bodies*, 276.

30. AHUNAM, F-ENM, R-ISM, S-MAP, caja 40, expediente 1, January 14, 1895, 2–3.

31. AHUNAM, F-ENM, R-ISM, S-MAP, caja 40, expediente 1, January 14, 1895, 3.

32. AHUNAM, F-ENM, R-ISM, S-MAP, caja 40, expediente 1, January 14, 1895, 4. Conversions are based on historic labor earnings using 1.92 pesos to the dollar for the year 1895. Unskilled labor accounts for the position of dissection room attendant; all other positions are based on skilled labor rates. Conversions for laboratory equipment and materials are based on historical commodity value.

33. Hardwicke, *Medical Education and Practice*, 4.

34. Hardwicke, 21.

35. Hardwicke, 135.

36. AHUNAM, F-ENM, R-ISM, S-MAP, caja 40, expediente 1, January 14, 1895, 4–5.

37. AHUNAM, F-ENM, R-ISM, S-MAP, caja 40, expediente 4, August 1896, 83. Conversions are based on historic labor earnings using 1.92 pesos to the dollar for the year 1895.

38. AHUNAM, F-ENM, R-ISM, S-MAP, caja 40, expediente 4, August 1896, 83. Conversions are based on historic labor earnings using 1.908 pesos to the dollar for the year 1896.

39. Bates, "'Indecent and Demoralizing Representations,'" 1.

40. AHUNAM, F-ENM, R-ISM, S-MAP, caja 40, expediente 1, January 14, 1895, 3.

41. AHUNAM, F-ENM, R-ISM, S-MAP, caja 40, expediente 4, December 1896, 82.

42. AHUNAM, F-ENM, R-D, Sub-Sec, S-Correspondencia General (hereafter S-CG), caja 22, expediente 44, April 12, 1897, 223.

43. Hardwicke, *Medical Education and Practice*, 1.

44. AHUNAM, F-ENM, R-D, Sub-Sec, S-CG, caja 22, expediente 44, April 27, 1897, 225.

45. AHUNAM, F-ENM, R-D, Sub-Sec, S-PE, caja 18, expediente 7, December 2, 1897, 79–80.

46. AHUNAM, F-ENM, R-D, Sub-Sec, S-PE, caja 18, expediente 7, December 2, 1897, 80–81.

47. AHUNAM, F-ENM, R-D, Sub-Sec, S-PE, caja 18, expediente 7, December 2, 1897, 82.

48. AHUNAM, F-ENM, R-D, Sub-Sec, S-PE, caja 18, expediente 7, December 2, 1897, 82–84. In the United States during the late nineteenth century, dissection had become a stepping stone for individuals interested in becoming medical professionals. Yet the physicians this system produced were less interested in dissecting bodies exclusively and more interested in using popular medical and criminological theories to argue that moral, social, or political categories were "natural" or "pathological." For more, see Geison, "'Divided We Stand'"; Shortt, "Physicians, Science, and Status"; Warner, "Science in Medicine"; Rothstein, *American Medical Schools*; Bynum, *Science and the Practice*; and Warner, "Histories of Science."

49. AHUNAM, F-ENM, R-D, Sub-Sec, S-PE, caja 18, expediente 7, December 2, 1897, 84.

50. AHUNAM, F-ENM, R-D, Sub-Sec, S-PE, caja 18, expediente 7, December 15, 1897, 109.

51. AHUNAM, F-ENM, R-D, Sub-Sec, S-PE, caja 18, expediente 7, May 27, 1898, 157.

52. AHUNAM, F-ENM, R-D, Sub-Sec, S-PE, caja 18, expediente 9, 1899, 374.

53. Starr, *Readings from Modern Mexican*, 359.

54. AHUNAM, F-ENM, R-D, Sub-Sec, S-PE, caja 18, expediente 7, September 26, 1898, 287.

55. AHUNAM, F-ENM, R-D, Sub-Sec, S-PE, caja 18, expediente 7, September 26, 1898, 288.

56. "Obituaries," 540.

57. AHUNAM, F-ENM, R-D, Sub-Sec, S-PE, caja 18, expediente 9, September 26, 1898, 287. The textbook that Francisco de P. Chacón had chosen was written by French physician Paul Jules Tillaux, who had also produced a publication about the surgical treatment of bone fractures.

58. Bartels and van Overbeeke, "Historical Vignette," 477.

59. AHUNAM, F-ENM, R-D, Sub-Sec, S-PE, caja 18, expediente 9, September 26, 1898, 287.

60. "Miscellany."

61. AHUNAM, F-ENM, R-D, Sub-Sec, S-PE, caja 18, expediente 9, October 1898, 305.

62. Scott, *Seeing Like a State*, 78.

63. AGN, TSJDF, caja 372, expediente 343, February 15, 1900, 1.

64. Hart, *Empire and Revolution*, 237; and "Railroading in Mexico," *The Railway Agent and Station Agent: A Monthly Journal Devote to the Interests of Local Freight and Freight Agents* 1 (March 1889): 172.

65. "Railroading in Mexico," *The Railway Agent*, 172.

66. Hart, *Empire and Revolution*, 238.

67. Rohé, "Resolutions Adopted," 5. Monjarás had achieved international recognition as a Mexican commissioner in the American Public Health Association along with physicians Eduardo Licéaga and José Ramírez.

68. AGN, F-TSJDF caja 373, expediente 343, May 26, 1900, 25–26.

69. AGN, F-TSJDF caja 373, expediente 343, May 26, 1900, 26.

70. AGN, F-TSJDF caja 373, expediente 343, May 26, 1900, 27.

71. AGN, F-TSJDF caja 373, expediente 343, May 1900, 15–16.

72. AGN, F-TSJDF caja 373, expediente 343, May 28, 1900, 15–17.

73. AGN, F-TSJDF caja 373, expediente 343, June 7, 1900, 7.

74. AGN, F-TSJDF caja 373, expediente 343, June 4, 1900, 19.

75. AGN, F-TSJDF caja 373, expediente 343, June 4, 1900, 19–20.

76. AGN, F-TSJDF caja 373, expediente 343, June 4, 1900, 20.

77. Agostoni, *Monuments of Progress*, 27.

78. AGN, F-TSJDF caja 373, expediente 343, June 4, 1900, 21–23.

79. AGN, F-TSJDF caja 373, expediente 343, June 4, 1900, 21–23.

80. AGN, F-TSJDF caja 373, expediente 343, June 7, 1900, 33.

81. Matthews, "Railway Culture," 177–79.

82. AGN, F-TSJDF caja 373, expediente 343, June 7, 1900, 36–37.

83. AGN, F-TSJDF caja 373, expediente 343, June 7, 1900, 38.

84. AGN, F-TSJDF caja 373, expediente 343, July 11, 1900, 40.

85. Pérez-Rayón, "La Sociología," 182.

86. Armus, "Tango, Gender, and Tuberculosis in Buenos Aires," 101–29.

87. Carrillo, "¿Estado de peste o estado de sitio?"

88. "Society Proceedings," *Buffalo Medical Journal* 36, no. 2 (August 1896–July 1897), 203.

89. AHCM, F-AM/GDF, S-P, caja 4, expediente 264, January 31, 1901, 1.

90. AHCM, F-AM/GDF, S-P, caja 4, expediente 309, May 14, 1901, 1.

91. AHUNAM, F-ENM, R-ISM, S-MAP, caja 40, expediente 13, May 1901, 262.

92. Piccato, *City of Suspects*, 34–41.

93. González Navarro, *Estadisticas Sociales del Porfiriato*, 24–25.

94. AHUNAM, F-ENM, R-ISM, S-MAP, caja 40, expediente 13, May 1901, 263–64. Interestingly, American physicians working in the 1920s and 1930s for Henry Ford's rubber plantation along the Tapajós River in the Amazon (Fordlandia) believed beriberi was caused by patients sleeping in low-lying hammocks with their backs too close to the cold clay floors of their homes. For more, see Grandin, *Fordlandia*, 273–74.

95. AHUNAM, F-ENM, R-ISM, S-MAP, caja 40, expediente 13, May 1901, 265.

96. For more, see Pilcher, *Sausage Rebellion*; Piccato, *City of Suspects*.

97. For a discussion of how state officials controlled citizen's bodies in both the United States and Sweden through science, see Sappol, *Traffic of Dead Bodies*, 274–309; Åhrén, *Death, Modernity*.

98. Voekel, "Peeing on the Palace," 183–208.

99. Piccato, *City of Suspects*, 29–30; Johns, *City of Mexico*, 5–6.

100. Toxqui Garay, "'El Recreo de Los Amigos'"; Mitchell, *Intoxicating Identities*; Lear, *Workers, Neighbors, and Citizens*, 89–142; Piccato, *City of Suspects*, 20–40; and Pilcher, *Sausage Rebellion*, 96–98.

101. Piccato, *City of Suspects*, 28–30.

102. Piccato, 70–71.

103. AHUNAM, F-ENM, R-D, Sub-Sec, S-PE, caja 19, expediente 13, August 5, 1901, 523.

104. AHUNAM, F-ENM, R-D, Sub-Sec, S-PE, caja 19, expediente 24, July 23, 1906, 829.

105. Giddens, *Modernity and Self-Identity*, 15–16.

106. AHUNAM, F-ENM, R-D, Sub-Sec, S-PE, caja 19, expediente 24, July 23, 1906, 831.

107. AHUNAM, F-ENM, R-D, Sub-Sec, S-PE, caja 19, expediente 24, July 23, 1906, 832–33. Conversions are based on historical opportunity costs using the year 1907.

108. Robbins, *Louis Pasteur*, 78–111.

109. Conversions are based on historical real price using 2.008 pesos to the dollar for the year 1907.

110. AHUNAM, F-ENM, R-D, Sub-Sec, S-PGI, caja 32, expediente 39, December 19, 1906, 207.

111. AHUNAM, F-ENM, R-D, Sub-Sec, S-PE, caja 19, expediente 24, July 23, 1906, 833–36. I calculated the total number of hours based on the program information for first- through fifth-year medical school. The number does not take into account any additional practice students received outside of sanctioned classes. For example, see the case of Guadalupe Rodríguez, a medical attendant at the National School of Medicine whom the university fired for selling cadavers to students. His case appears herein in chapter 4 and in AHUNAM, F-ENM, R-D, Sub-Sec, S-CG, caja 23, expediente 67, September 8, 1905, 444–50.

112. Licéaga, *Memoria de Los Trabajos*, 4–7. For more on the university's relationship with the Mexican government, see Sánchez, "La Universidad Nacional."

113. Licéaga, *Memoria de Los Trabajos*, 4–7.

114. Licéaga, 6–7.

115. Agostoni, *Monuments of Progress*, 76.

116. Licéaga, *Memoria de Los Trabajos*, 8.

117. Scott, *Seeing Like a State*, 92.

118. See Voekel, *Alone before God*, 123–219; Wasserman, *Everyday Life and Politics*, 99–158; Powell, "Priest and Peasants."

119. Overmeyer-Velázquez, *Visions*, 70–97.

120. Kapelusz-Poppi, "Provincial Intellectuals from Michoacan," 7. For more on the role of physicians during the reign of Porfirio Díaz, see Carrillo, "Profesiones sanitarias"; Agostoni, "Popular Health Education," 52–61.

121. AHUNAM, F-ENM, R-D, Sub-Sec, Serie-Informes y Memorias (hereafter S-IM), caja 30, expediente 6, May 2, 1910, 7.

122. González Navarro, *Estadisticas Sociales del Porfiriato*, 19.

123. Anderson, *Outcasts in Their Own Land*, 299.

124. AHUNAM, F-ENM, R-D, Sub-Sec, S-CG, caja 25, expediente 113, 1912, 1127–39.

125. AHUNAM, F-ENM, R-D, Sub-Sec, S-PE, caja 2, expediente 21542, June 18, 1918, 3; AHUNAM, F-ENM, R-D, Sub-Sec, S-PE, caja 2, expediente 21542, July 19, 1918, 15; and "El Día de Cadáver Anónimo," *El Pueblo* July 30, 1918, 3.

126. AHUNAM, F-ENM, R-D, Sub-Sec, S-PE, caja 2, expediente 21542, July 19, 1918, 15.

127. "El Día de Cadáver Anónimo," *El Diario Ilustrado de la Mañana* July 31, 1918, 11. For more on photographs that captured these moments at medical schools in the United States, see Warner and Edmonson, *Dissection*.

128. AHUNAM, R-D, Sub-Sec, Series-Facultad de Medicina Expedientes de Personal (hereafter S-FMEP), caja 10, expediente 4705, October 6, 1920, 73.

129. AHCM, F-SP, Sección-Manicomio General (hereafter Sec-MG), caja 8, expediente 5, November 1, 1942, 14–15; AHSSA, F-SP, Sec-MG, caja 8, expediente 5, October 11, 1944, 28–29; AHSSA, F-SP, Sec-MG, caja 8, expediente 5, August 12, 1954, 56.

3. Wet or Dry Remains

1. Beatty and Sáiz González, "Industrial Property Institutions," 3. For more on the history of technology transfer, see David, "Path Dependence," 15–36; O'-Rourke and Wilson, *Globalization and History*; Edgerton, "From Innovation to Use"; Söderquist, *Historiography of Contemporary Science*; Mokyr, *Levers of Riches*; Jeremy, *Transatlantic Industrial Revolution*; Allen, Tushman, and Lee, "Technology Transfer"; Rosenberg, "Economic Development and the Transfer of Technology"; and Gerschenkron, *Economic Backwardness*.

2. Beatty, *Technology and the Search*, 3.

3. Haber, Razo, and Maurer, *Politics of Property Rights*, 47.

4. Edward Beatty, email message to the author, October 24, 2012. I determined this number based on records provided to me by Professor Beatty of patents he has collected that are missing from the records of Grupo Documental-Patentes y Marcas in the Archivo General de la Nación.

5. Agostoni, *Monuments of Progress*; Piccato, *City of Suspects*; Bazant, "La enseñanza."

6. Branes, *Patent Technology*.

7. Beatty and Sáiz González, "Industrial Property Institutions," 18. Conversions based on contemporary opportunity costs using 2.375 pesos to the dollar for the year 1903.

8. Edward Beatty, email message to the author, July 25, 2012.

9. Beatty and Sáiz González, "Industrial Property Institutions," 12.

10. Khan, *Democratization of Invention*, 54.

11. Sandwick, "More Men in Public School," 450. Conversions based on historical labor earnings (unskilled) using the year 1890.

12. Piccato, *City of Suspects*, 246–47. Conversions based on historical labor earnings using 2.062 pesos to the dollar for the year 1900. Assuming that the wage laborer worked every single day at the rate of one peso per day, his annual income would be 365 pesos or $24,820 in 2016. However, this is a false equivalency since most laborers did not work consistently during the year.

13. Habenstein and Lamers, *American Funeral Directing*, 158–60.

14. Hohenschuh, *Modern Funeral*, 9.

15. Farrell, *Inventing the American Way*, 171.

16. Habenstein and Lamers, *American Funeral Directing*, 164.

17. Habenstein and Lamers, 167.

18. AGN, Grupo Documental-Patentes y Marcas (hereafter GD-PM), legajo 307, expediente 84, November 22, 1907, 1.

19. AGN, GD-PM, legajo 307, expediente 84, November 22, 1907, 1.

20. Esposito, *Funerals, Festivals*, 133.

21. AGN, GD-PM, legajo 307, expediente 86, May 28, 1910, 1.

22. While there are no definitive sources that state whether or not body-snatching did occur in Mexico, a close examination of hundreds of cemetery records for Mexico City have indicated that corpses were not stolen from graves. There are, however, several cases that illustrate that theft of items placed on top of an individual's grave, such as small statues, jewelry, nickel-plated crosses or flower pots, was quite common. For more, see AHCM, F-AM/GDF, S-P, caja 3497, expediente 761, September 30, 1911, 1–2; AHCM, F-AM/GDF, Serie-Panteón Dolores (hereafter S-PD), caja 3483, expediente 251, November 24, 1905, 1; AHCM, F-AM/GDF, S-PD, caja 349, expediente 736, March 23, 1912, 1; AHCM, F-AM/GDF, S-PD, caja 349, expediente 736, April 12, 1912, 5. According to historian Amanda López, it was common at this time to reuse markers and monuments. The cemetery had a bodega in which they offered used monuments for sale. For more, see López, "Cadaverous City," 65.

23. Tulchinsky and Varivakova, *New Public Health*, 14–16. For more, see Whooley, *Knowledge in the Time*, López-Alonso, *Measuring Up*, Ewald, *Plague Time*, and Rosenberg, *Cholera Years*.

24. Voekel, *Alone before God*, 172–219.

25. Esposito, *Funerals, Festivals*, 45–51.

26. John Charles Fremont McGriff, Burial Casket, U.S. Patent 684,293, filed on October 17, 1900, and issued on October 8, 1901.

27. AGN, GD-PM, legajo 307, expediente 74, October 1, 1903, 1.

28. AGN, GD-PM, legajo 307, expediente 74, October 1, 1903, 1.

29. AGN, GD-PM, legajo 307, expediente 74, October 1, 1903, 1–2.

30. Cabrejo, "New Order," 163. A sense of modesty was an attribute shared by many Latin American elites across the region during the late nineteenth and early twentieth century. For more on how elites in Mexico used the concept to separate themselves from the lower classes, see Kelly, "Into the Galactic Zone," 31–44; French, *Peaceful and Working People*; Overmeyer-Velázquez, *Visions*; and Bliss, *Compromised Positions*.

31. AGN, GD-PM, legajo 307, expediente 79, October 1, 1903, 1.

32. AGN, GD-PM, legajo 307, expediente 79, October 1, 1903, 1.

33. Mitford, *American Way of Death*, 84.

34. AGN, GD-PM, legajo 307, expediente 78, April 26, 1904, 1.

35. Lavery, *Horatio Lord Nelson*, 6.

36. AGN, GD-PM, legajo 307, expediente 80, March 1, 1907, 1.

37. Helen Conger, email message to Case Western Archivist, August 1, 2012.

38. Campbell, *Biographical History*, 361.

39. At https://lloydlibrary.org/wp-content/themes/Lloyd_wptheme/finding-aids/Eclectic%20Medical%20Institute%20Matriculation%20Roster-1845-1939--composed%20by%20John%20Haller.pdf you can find the most recent information on the Eclectic Medical Institute and its available matriculation records, as of 2018. For more on the Eclectic Medical School, see Haller, *Medical Protestants*; and Sappol, *Traffic of Dead Bodies*, 137–52. For more on congestive intermittent, see Hunt, *Homeopathic Theory and Practice*, 510.

40. Campbell, *Biographical History*, 361.

41. Fischer, *Nephritis*, 2.

42. "Monkey: A Suicide through Grief," *Free-Lance Star*, July 24, 1909, 2.

43. Monroe S. Leech, U.S. patent number 809,573, filed on August 21, 1905, and issued on January 9, 1906; he had also received a patent in the United States for a method of preserving human bodies, patent number 826,583, filed on February 6, 1906, and issued on July 24, 1906.

44. Habenstein and Lamers, *American Funeral Directing*, 178.

45. AGN, GD-PM, legajo 307, expediente 80, March 1, 1907, 1.

46. AGN, GD-PM, legajo 307, expediente 80, March 1, 1907, 1.

47. AGN, GD-PM, legajo 307, expediente 83, September 5, 1911, 1.

48. Elijah D. McDonald, Trolley, U.S. Patent Number 833,080, filed on August 31, 1905, and issued on October 9, 1906; Elijah D. McDonald, Self-lubricating Wheel, U.S. Patent Number 832,994, filed on August 31, 1905, and issued on October 9, 1906; Elijah D. McDonald, New Form of Railway Construction, U.S. Patent Number 943,198, filed on December 12, 1906, and issued on December 14, 1909; Elijah D. McDonald, Reinforced Concrete Pile, U.S. Patent Number 979,529, filed on January 13, 1910, and issued on December 27, 1910; and Elijah D. McDonald, Mortuary Box, U.S. Patent Number 1,168,660, filed on April 29, 1912, and issued on January 18, 1916.

49. AGN, GD-PM, legajo 307, expediente 83, September 5, 1911, 1–2.

50. Millet, *Roman Britain*, 128; and Adkins and Adkins, *Handbook to Life*, 357.

51. Cheesman, *Environmental Impacts of Sugar*, 161–64.

52. Iserson, *Death to Dust*, 194.

53. Urbain, *La Societe de Conservation*, 67–68.

54. Habenstein and Lamers, *American Funeral Directing*, 183–84.

55. AGN, GD-PM, legajo 207, expediente 95, May 6, 1912, 1.

56. AGN, GD-PM, legajo 207, expediente 95, May 6, 1912, 1.

57. AGN, GD-PM, legajo 207, expediente 96, June 19, 1912, 1. For more on the spatial separation of elite families in Mexico City during the Porfiriato, see Piccato, *City of Suspects*, 20–33.

58. AGN, GD-PM, legajo 207, expediente 99, July 24, 1913, 1.

59. Ashbee, *Earthen Long Barrow.*

60. Habenstein and Lamers, *American Funeral Directing*, 200–201.

61. Farrell, *Inventing the American Way*, 159.

62. AGN, GD-PM, legajo 207, expediente 76, October 1, 1903, 1.

63. AGN, GD-PM, legajo 207, expediente 76, October 1, 1903, 1.

64. AGN, GD-PM, legajo 207, expediente 79, February 17, 1906, 1.

65. AGN, GD-PM, legajo 207, expediente 79, February 17, 1906, 2. The use of silver nitrate made for an interesting choice because of its naturally occurring antimicrobial properties, though Johnston himself was mostly likely unaware of this feature. Silver nitrate also stained the body, leaving dark spots wherever poured on the skin.

66. Chris Quigley, "Electroplating the Dead," March 17, 2011, http://www.quigleyscabinet.blogspot.com/2011/03/electroplating-dead.html.

67. Habenstein and Lamers, *American Funeral Directing*, 201.

68. Levon G. Kassabian, Method of Preserving Dead Bodies, U.S. Patent 2,023,685, filed on February 2, 1934, and issued on December 10, 1935.

69. AGN, GD-PM, legajo 207, expediente 81, May 3, 1907, 1.

70. AGN, GD-PM, legajo 207, expediente 81, May 3, 1907, 2.

71. AGN, GD-PM, legajo 207, expediente 82, August 13, 1907, 1.

72. AGN, GD-PM, legajo 207, expediente 82, August 13, 1907, 1–2. By 1892 many medical professionals were familiar with the microbicidal properties of formaldehyde. This led to the development of a solution that German doctors Hans Rosemann and Alfred Stephan called Lysoform, made from liquid soap and formaldehyde, which they submitted for patenting in 1900 to the German patent office. For more, see http://www.lysoform.com/lyso.htm, which outlines the history of Lysoform from its parent company, last updated in 2011.

73. AGN, GD-PM, legajo 207, expediente 97, September 13, 1912, 1.

74. Iserson, *Death to Dust*, 192.

75. Mayer, *Embalming*, 23–57.

76. Iserson, *Death to Dust*, 193–94. For more on Robert Boyle, see Hunter, *Robert Boyle*; and Hunter, *Robert Boyle Reconsidered.*

77. J. A. Gaussardia, Preserving Dead Bodies, U.S. Patent 15,972, issued on October 28, 1856.

78. E. B. White and Ivan Sandof, "The First Embalmer," *New Yorker*, November 7, 1942, 44. In 1900 2.06 pesos equaled 1 U.S. dollar. German chemist August Wilhelm von Hofman discovered formaldehyde in 1867, but scientists and physicians did not realize its potential as a preservative until 1888; it would be several years before the use of formaldehyde became commonplace among embalming professionals due to its high cost.

79. Habenstein and Lamers, *American Funeral Directing*, 224–25.

80. Habenstein and Lamers, 228–30.

81. White and Sandof, "First Embalmer," 44.

82. AGN, GD-PM, legajo 207, expediente 85, May 29, 1908, 1.

83. AHCM, F-AM/GDF, S-P, caja 3475, expediente 28, July 7, 1918, 4.

84. Knight, *Mexican Revolution*, 1:40.

85. AHCM, F-AM/GDF, S-P, caja 3475, expediente 28, July 7, 1918, 4–5.

86. AHCM, F-AM/GDF, S-P, caja 3475, expediente 28, July 7, 1918, 4–5.

87. AHCM, F-AM/GDF, S-P, caja 3475, expediente 28, July 7, 1918, 7–8.

88. AHCM, F-AM/GDF, S-P, caja 3475, expediente 28, July 7, 1918, 9–10.

89. AHCM, F-AM/GDF, S-P, caja 3475, expediente 28, July 7, 1918, 11–13.

90. AHCM, F-AM/GDF, S-P, caja 3475, expediente 28, July 7, 1918, 13.

91. AHCM, F-AM/GDF, S-P, caja 3475, expediente 28, July 7, 1918, 10–12. For more examples of lower-class professions that worked alongside the dead, see López, "The Cadaverous City," 74–80.

92. AHCM, F-AM/GDF, S-P, caja 3475, expediente 28, November 6, 1918, 1.

93. Crosby, *America's Forgotten Pandemic*, 5.

94. For more information on the controversy surrounding mortality statistics during the Spanish Influenza, see Knight, *Mexican Revolution*, 2:422; and Chowell, Viboud, Simonsen, Miller, and Acuna-Soto, "Mortality Patterns."

95. AHCM, F-AM/GDF, S-P, caja 3475, expediente 28, November 6, 1918, 1.

96. Conversions are based on historic labor earnings (a range of unskilled to skilled labor) using 1.807 pesos to the dollar for the year 1918.

97. AHCM, F-AM/GDF, S-P, caja 3475, expediente 28, November 6, 1918, 2.

98. AHCM, F-AM/GDF, S-P, caja 3475, expediente 28, November 19, 1918, 1.

99. Unlike early twentieth-century Buenos Aires, London, Paris, or San Francisco, where civic leaders popularized and financed cremation, the Mexican government was responsible for the construction of the crematory ovens. For more, see López, "Cadaverous City," 83–85.

100. AHCM, F-AM/GDF, S-P, caja 3486, expediente 446, October 12, 1907, 1–2. Conversions are based on contemporary opportunity costs using 2.011 pesos to the dollar for the year 1908.

101. AHCM, F-AM/GDF, S-P, caja 3486, expediente 446, October 12, 1907, 3–4.

102. AHCM, F-AM/GDF, S-P, caja 3486, expediente 460, December 5, 1907, 1.

103. López, "Cadaverous City," 56. Conversions are based on contemporary opportunity costs using 2.062 pesos to the dollar for the year 1900, when contemporary opportunity cost could first be calculated.

104. AHCM, F-AM/GDF, S-P, caja 3472, expediente 257, November 11, 1919, 5. Conversions based on contemporary opportunity costs using 1.985 pesos to the dollar for the year 1919.

105. López, "Cadaverous City," 30.

106. AHCM, F-AM/GDF, S-P, caja 3486, expediente 460, December 5, 1907, 1–2.

107. AHCM, F-AM/GDF, S-P, caja 3486, expediente 460, December 5, 1907, 1–2.

108. Agostoni, *Monuments of Progress*, 23–24.

109. López, "Cadaverous City," 83.

110. Feldenkirchen, *Werner Von Siemens*, 13.

111. Buklijas, "Culture of Death," 476.

112. Lomnitz, *Death and The Idea*, 159–72.

113. Esposito, *Funerals, Festivals*, 113–44.

114. "The Church and Cremation: Why Was It Forbidden?" *The Remnant*, May 15, 2007.

115. Iserson, *Death to Dust*, 275; Farrell, *Inventing the American Way*, 166. On May 8, 1963, the Roman Catholic Church issued an instruction titled "Piam et constantem," which hesitantly permitted cremation. According to the church, its opposition to cremation was done to prevent "hate-inspired" attacks against the Christian practices and traditions (i.e., burials). For more, see Suprema Sacra Congregatio Sancti Officii, *Instructio Piam et constantem: de cadaverum cremation*, in *Acta Apostolicae Sedis* 56 (1964), 822–24.

116. "Denouncing Cremation: A Decree from Rome—The Feast of St. Alphonso's," *New York Times*, August 3, 1886; Porro et al., "Modernity in Medicine."

117. Prothero, *Purified by Fire*, 157.

118. López, "Cadaverous City," 89–94.

119. Vega and Baez, "La Óptica Metodista," 11.

120. Borton, "¿Hay razones de orden religiosa para no incinerar nuestro cadáveres?," *El Abogado Cristiano Ilustrado*, July 15, 1909, 437.

121. Garza, "¿Es la incineración de cadáveres anticristiano?," *El Abogado Cristiano Ilustrado*, July 15, 1909, 438.

122. Borton,"¿Hay razones?," 437.

123. Farrell, *Inventing the American Way*, 166–67.

124. Ariès, *Western Attitudes toward Death*, 91.

125. Laura Gómez Flores, "Diabetes mantiene en emergencia al país: Mancera," *La Jornada en linea*, November 14, 2016, http://www.jornada.unam.mx/ultimas/2016/11/14/diabetes-mantiene-en-emergencia-al-pais-mancera; Ángeles Cruz Martínez, "México, en emergencia epidemiológica por diabetes: SSA," *La Jornada en linea*, November 15, 2016, http://www.jornada.unam.mx/ultimas/2016/11/14/mexico-en-emergencia-epidemiologica-por-sobrepeso-ssa.

126. Jacqueline Howard, "Do Soda Taxes Work?: Experts Look to Mexico for Answers," CNN, November 1, 2016, http://www.cnn.com/2016/11/01/health/soda-tax-benefits-mexico/.

127. Amy Guthrie and Mike Esterl, "Soda Sales in Mexico Rise Despite Tax," *Wall Street Journal*, May 3, 2016, http://www.wsj.com/articles/soda-sales-in-mexico-rise-despite-tax-1462267808.

128. AGN, GD-PM, legajo 307, expediente 75, December 2, 1925, 1.

129. AGN, GD-PM, legajo 307, expediente 75, December 2, 1925, 1.

130. AGN, GD-PM, legajo 307, expediente 75, December 2, 1925, 2.

4. Undermining Progress

1. AGN, TSJDF, caja 470, expediente 82690, March 15, 1906, 1.

2. AGN, TSJDF, caja 470, expediente 82690, March 15, 1906, 3.

3. AGN, TSJDF, caja 1194, expediente 210889, May 7, 1913, 1; AHCM, F-AM/GDF, S-P, caja 4, expediente 258, February 22, 1901; AHCM, F-AM/GDF, S-P, caja 6, expediente 498, January 27, 1902.

4. AHCM, F-AM/GDF, S-P, caja 2, expediente 272, December 31, 1900, 1.

5. Agostoni, *Monuments of Progress*, 68–73.

6. Piccato, *City of Suspects*, 48.

7. Agostoni, *Monuments of Progress*, 57.

8. AHCM, F-AM/GDF, S-P, caja 3458, expediente 575, February 10, 1893, 1.

9. López, "Cadaverous City," 64–65.

10. AHCM, F-AM/GDF, S-P, caja 3458, expediente 575, 10 February 1893, 1.

11. AHCM, F-AM/GDF, S-P, caja 3458, expediente 575, February 10, 1893, 2.

12. AHCM, F-AM/GDF, S-P, caja 3458, expediente 575, February 22, 1893, 1–2.

13. Beezley, *Judas at the Jockey Club*, 73–74.

14. López, "The Cadaverous City," 74–75.

15. Lomnitz, *Death and the Idea*, 170–77; Esposito, *Funerals, Festivals*, 27; Megged, *Social Memory*, 136–40.

16. Piccato, *City of Suspects*, 34–49.

17. Voekel, *Alone before God*, 106–45.

18. Esposito, *Funerals, Festivals*, 1–19. For a firsthand account of Day of the Dead celebrations from the early twentieth century, see Tweedie, *Mexico As I Saw It*, 215–17. For more on the Day of the Dead, see Brandes, *Skulls to the Living*.

19. Lomnitz, *Death and the Idea*, 329.

20. Esposito, *Funerals, Festivals*, 39.

21. Lomnitz, *Death and the Idea*, 330–31.

22. AHCM, F-M/GDF, S-P, caja 5, expediente 359, August 23, 1901, 1–3. Conversions based on contemporary opportunity cost using 2.062 pesos to the dollar for the year 1900.

23. Piccato, *City of Suspects*, 246–47.

24. Terry, *Terry's Mexico*, 232–33.

25. AHCM, F-AM/GDF, S-P, caja 5, expediente 359, August 23, 1901, 1.

26. Esposito, *Funerals, Festivals*, 23–35.

27. Conversions based on contemporary opportunity cost using 2.114 pesos to the dollar for the year 1901.

28. AHCM, F-AM/GDF, S-P, caja 5, expediente 359, August 30, 1901, 3.

29. AHCM, F-AM/GDF, S-P, caja 5, expediente 359, August 31, 1901, 5.

30. AHCM, F-AM/GDF, S-P, caja 5, expediente 359, August 31, 1901, 6.

31. Scott, *Moral Economy*, 4–7.

32. AHCM, F-AM/GDF, S-P, caja 5, expediente 359, August 31, 1901, 6.

33. AHCM, F-AM/GDF, S-P, caja 5, expediente 359, September 12, 1901, 7.

34. Terry, *Terry's Mexico*, 400.

35. AHCM, F-AM/GDF, S-P, caja 6, expediente 498, April 14, 1902, 1.

36. AHCM, F-AM/GDF, S-P, caja 6, expediente 498, April 18, 1902, 4. Conversions based on historic opportunity cost using 2.387 pesos to the dollar for the year 1902.

37. AHCM, F-AM/GDF, S-P, caja 6, expediente 498, April 28, 1902, 7.

38. AHCM, F-AM/GDF, S-P, caja 6, expediente 498, May 15, 1902, 6.

39. AHCM, F-AM/GDF, S-P, caja 4, expediente 264, January 31, 1901, 1; AHCM, F-AM/GDF, S-P, caja 4, expediente 264, February 5, 1901, 4; AHCM, F-AM/GDF, S-P, caja 4, expediente 309, May 14, 1901, 1–2.

40. Scott, *Seeing Like a State*, 28–47.

41. Iglesia, *Columbus, Cortés*, 151–54.

42. AHCM, F-AM/GDF, S-P, caja 9, expediente 896, January 28, 1903, 2.

43. AHCM, F-AM/GDF, S-P, caja 9, expediente 896, January 28, 1903, 2.

44. AHCM, F-AM/GDF, S-P, caja 9, expediente 896, February 7, 1903, 1–2.

45. AHCM, F-AM/GDF, S-P, caja 9, expediente 896, February 11, 1903, 3. In fact, the 1895 Sanitary Code contains a reference table for readers to see what article numbers had changed from the code's earlier versions. The book lists article 239 as article 203 in the 1895 version of the sanitary code. For the original Spanish version of the article, see Article 203 in *Codigo Sanitario*, 49.

46. AHCM, F-AM/GDF, S-P, caja 9, expediente 896, February 11, 1903, 3. The SSC eventually changed the length of short-term burials in Ixtapalapa from five years to seven years on August 17, 1903. See AHCM, F-AM/GDF, S-P, caja 9, expediente 896, August 17, 1903, 5.

47. AHCM, F-AM/GDF, S-P, caja 9, expediente 861, June 5, 1903, 1.

48. AHCM, F-AM/GDF, S-P, caja 9, expediente 861, June 5, 1903, 1.

49. AHCM, F-AM/GDF, S-P, caja 9, expediente 861, June 5, 1903, 1.

50. Agostoni, *Monuments of Progress*, 28–30.

51. French, *Peaceful and Working People*, 63.

52. AHCM, F-AM/GDF, S-P, caja 6, expediente 576, June 2, 1902, 1.

53. AHCM, F-AM/GDF, S-P, caja 6, expediente 576, June 9, 1902, 6.

54. AHCM, F-AM/GDF, S-P, caja 6, expediente 576, June 9, 1902, 6–7.

55. AHCM, F-AM/GDF, S-P, caja 6, expediente 576, June 9, 1902, 8.

56. AHCM, F-AM/GDF, S-P, caja 6, expediente 576, June 9, 1902, 10–11.

57. Lomnitz, *Death and the Idea*, 35–41.

58. Taylor, *Buried Soul*, 268.

59. Scott, *Seeing Like a State*, 91–92.

60. Johns, *City of Mexico*, 81–82; and Lomnitz, *Death and the Idea*, 322.

61. Terry, *Terry's Mexico*, 419.

62. Terry, 419–20.

63. AGN, TSJDF, caja 358, expediente 57353, January 29, 1904, 1–2.

64. AGN, TSJDF, caja 358, expediente 57353, January 29, 1904, 3–5.

65. AGN, TSJDF, caja 358, expediente 57353, January 29, 1904, 8.

66. AGN, TSJDF, caja 358, expediente 57353, January 29, 1904, 9.

67. Hale, *Transformation of Liberalism*, 206–42; and Voekel, *Alone before God*, 221–22.

68. AHUNAM, F-ENM, R-D, Sub-Sec, S-CG, caja 23, expediente 67, August 31, 1905, 445.

69. AHUNAM, F-ENM, R-D, Sub-Sec, S-CG, caja 23, expediente 67, August 31, 1905, 445.

70. AHUNAM, F-ENM, R-D, Sub-Sec, S-CG, caja 23, expediente 67, September 8, 1905, 447.

71. AHUNAM, F-ENM, R-D, Sub-Sec, S-CG, caja 23, expediente 67, September 8, 1905, 447.

72. AHUNAM, F-ENM, R-D, Sub-Sec, S-CG, caja 23, expediente 67, September 8, 1905, 448.

73. Piccato, *City of Suspects*, 9. Conversions based on unskilled labor earnings cost using 2.018 pesos the dollar for the year 1905. For more works dealing with popular resistance in Latin America against state imposed modernization, see Buffington, *Criminal and Citizen*; Johnson, *Problem of Order*; and Aguirre, *Criminals of Lima*.

74. AHUNAM, F-ENM, R-D, Sub-Sec, S-CG, caja 23, expediente 67, September 8, 1905, 448.

75. Cruz Gómez, "El Hospital Juárez."

76. "Need for Reform at the Juárez Morgue; Dead Room is a Disgrace; Government Proposes to Eradicate Evils," *Mexican Herald*, June 10, 1909, 1.

77. "Need for Reform," 1.

78. AGN, TSJDF, caja 1333, expediente 33533, June 20, 1916, 1–4.

79. For more about the Basilica and its importance to Mexican religious folklore and identity, see Brading, *Mexican Phoenix*.

80. Scott, *Moral Economy*, 43–44.

81. Taylor, *Buried Soul*, 267.

82. Scott, *Moral Economy*, 231.

Conclusion

1. Tweedie, *Porfirio Díaz*, 316.

2. Tweedie, 321–22.

3. López Ramos, *History of the Air*, 78–99.

4. Esposito, *Funerals, Festivals*; López, "Cadaverous City"; French, *Peaceful and Working People*; Overmeyer-Velázquez, *Visions*; Agostoni, *Monuments of Progress*; Piccato, *City of Suspects*; Bliss, *Compromised Positions*; Buffington, *Criminal and Citizen*; and Rivera Garza, "Masters of the Street."

5. Latour, *We Have Never Been*, 131.

BIBLIOGRAPHY

Archival and Manuscript Materials

Archivo General de la Nación (AGN)
  Patentes y Marcas (PM)
  Tribunal Superior de Justicia del Distrito Federal (TSJDF)
Archivo Histórico de la Ciudad de México (AHCM)
  Ayuntamiento de México/Gobierno del Distrito Federal (AM/GDF)
Archivo Histórico de la Secretaría de Salubridad (AHSSA)
  Salubridad Pública (SP)
Archivo Histórico de la Universidad Nacional Autónoma de México (AHUNAM)
  Escuela Nacional de Medicina (ENM)

Published Sources

Abel, Christopher. *Hygiene and Sanitation in Latin America, c. 1870–1950*. London: Institute of Latin American Studies, University of London, 1996.

Adams, John A., Jr. *Mexican Banking and Investment in Transition*. Westport CT: Quorum, 1997.

Addas, Michael. *Machines as the Measure of Man: Science, Technology, and Ideologies of Western Dominance*. Ithaca NY: Cornell University Press, 1989.

Adkins, Lesley, and Roy A. Adkins. *Handbook to Life in Ancient Rome*. New York: Oxford Press, 1998.

Agostoni, Claudia. *Monuments of Progress: Modernization and Public Health in Mexico City, 1876–1910*. Boulder: University Press of Colorado, 2003.

———. "Popular Health Education and Propaganda: In Times of Peace and War, 1890–1920." *American Journal of Public Health* 96, no. 1 (January 2006): 52–61.

———. "'Que no traigan al médico': Los profesionales de la salúd entre la crítica y la sátira (ciudad de México, siglos xix–xx)." In *Actores, espacios y debates en la historia de la esfera pública en la ciudad de México*, edited by Cristina Sacristán and Pablo Piccato, 97–120. Mexico City: Instituto de Mora, 2005.

Aguirre, Carlos. *The Criminals of Lima and Their Worlds: The Prison Experience, 1850–1935*. Durham NC: Duke University Press, 2005.

Åhrén, Eva. *Death, Modernity, and the Body in Sweden, 1880 to 1940*. Rochester NY: University of Rochester Press, 2008.

Alcaraz Hernández, Sonia. "Las pestilentes 'mansiones de la muerte': Los cementerios de la ciudad de México, 1870–1890." *Trace* 58 (December 2010): 95.

Alegre-Díaz, Jesus, William Herrington, Malaquías López-Cervantes, Louisa Gnatiuc, Raul Ramirez, Michael Hill, Colin Baigent, Mark I. McCarthy, Sarah Lewington, Rory Collins, Gary Whitlock, and Roberto Tapia-Conyer "Diabetes and Cause-Specific Mortality in Mexico City." *New England Journal of Medicine* 375 (2016): 1961–71.

Alexander, Anna Rose. *City on Fire: Technology, Social Change, and the Hazards of Progress in Mexico City, 1860–1910*. Pittsburgh PA: University of Pittsburgh Press, 2016.

Allen, Thomas J., Michael L. Tushman, and Denis M. S. Lee. "Technology Transfer as a Function of Position in the Spectrum from Research through Development to Technical Services." *Academy of Management Journal* 22, no. 4 (December 1979): 694–708.

Anderson, Rodney D. *Outcasts in Their Own Land: Mexican Industrial Workers, 1906–1911*. DeKalb: Northern Illinois University Press, 1976.

Anderson, Warwick. *The Cultivation of Whiteness: Science, Health, and Racial Destiny in Australia*. Melbourne, Australia: Melbourne University Press, 2002.

Andrews, George Reid. *The Afro-Argentines of Buenos Aires, 1800–1900*. Madison: University of Wisconsin Press, 1980.

———. *Blacks and Whites in São Paulo, Brazil, 1888–1988*. Madison: University of Wisconsin, 1991.

Ariès, Phillipe. *Western Attitudes toward Death: From the Middle Ages to the Present.* Baltimore: Johns Hopkins University Press, 1974.

Armus, Diego. *The Ailing City: Health, Tuberculosis, and Culture in Buenos Aires, 1870–1950*. Durham NC: Duke University Press, 2011.

———. "Tango, Gender, and Tuberculosis in Buenos Aires, 1900–1940." In *Disease in the History of Modern Latin America: From Malaria to AIDS*, edited by Diego Armus, 101–29. Durham NC: Duke University Press, 2003.

Arnold, David. *Colonizing the Body: State Medicine and Epidemic Disease in Nineteenth Century India*. Berkeley: University of California Press, 1993.

Arrom, Silvia. *Containing the Poor: The Mexico City Poor House, 1774–1871*. Durham NC: Duke University Press, 2000.

Ashbee, Paul. *The Earthen Long Barrow in Britain: An Introduction to the Study of Funeral Practice and Culture of Neolithic People of the Third Millennium B.C.* Toronto ON: University of Toronto Press, 1970.

Baker, Henry B. *Sanitation in 1890*. Chicago: Office of the Association, 1891.

Barnard, Henry. *School Architecture: Or Contributions to the Improvements of School-Houses*. New York: A. S. Barnes, 1849.

Bartels, Ronald H. M. A., and Jacobus J. van Overbeeke. "Historical Vignette: Charles Labbé." *Journal of Neurosurgery* 87 (September 1997): 477–80.

Bates, A. W. "'Indecent and Demoralizing Representations': Public Anatomy Museums in Mid-Victorian London." *Medical History* 52, no. 1 (2008): 1–22.

Bazant, Mílada. "La enseñanza y la práctica de la ingeniería durante el Porfiriato." *Historia Mexicana* 3 (1984): 254–97.

Beatty, Edward. *Technology and the Search for Progress in Modern Mexico*. Oakland: University of California Press, 2015.

Beatty, Edward, and Patricio Sáiz González. "Industrial Property Institutions, Patenting and Technology Investment in Spain and Mexico, c. 1820–1914." *Working Papers in Economic History No. 2007/02*, Universidad Autónoma de Madrid.

———. *Institutions and Investment: The Political Basis of Industrialization in Mexico before 1911*. Palo Alto CA: Stanford University Press, 2001.

Beezley, William H. *Judas at the Jockey Club and Other Episodes of Porfirian Mexico*. Lincoln: University of Nebraska, 1989.

———. *Mexico in World History*. New York: Oxford University Press, 2011.

Bernstein, Andrew. *Modern Passings: Death Rites, Politics, and Social Change in Imperial Japan*. Honolulu: University of Hawaii Press, 2006.

Birn, Anne-Emanuelle, and Ana María Carrillo. "Neighbours on Notice: National and Imperialist Interests in the American Public Health Association, 1872–1921." *Canadian Bulletin of Medical History* 25, no. 1 (2008): 225–54.

Bliss, Katherine E. *Compromised Positions: Prostitution, Public Health, and Gender Politics in Revolutionary Mexico City*. University Park: Pennsylvania State University Press, 2001.

Blum, Ann S. *Domestic Economies: Family, Work, and Welfare in Mexico City, 1884–1943*. Lincoln: University of Nebraska Press, 2009.

Boone, Christopher G. "Streetcars and Politics in Rio de Janeiro: Private Enterprise versus Municipal Government in the Provision of Mass Transit, 1903–1920." *Journal of Latin American Studies* 27, no. 2 (May 1995): 343–65.

Borton, F. S. "¿Hay razones orden religioso para no incinerar nuestros cadáveres?" El Abogado Cristiano Ilustrado. July 15, 1909.

Brading, D. A. *Mexican Phoenix: Our Lady of Guadalupe: Image and Tradition across Five Centuries*. New York: Cambridge University, 2001.

Brandes, Stanley. *Skulls to the Living, Bread to the Dead: The Day of the Dead in Mexico and Beyond*. Malden MA: Blackwell, 2006.

Branes, Juanita M., ed. *Patent Technology: Transfer and Industrial Competition*. New York: Nova Science Publishers, 2007.

Buffington, Robert. *Criminal and Citizen in Modern Mexico*. Lincoln: University of Nebraska Press, 2000.

Buffington, Robert, and William E. French. "The Culture of Modernity." In *The Oxford History of Mexico*, edited by William H. Beezley and Michael C. Meyer, 397–432. New York: Oxford University Press, 2010.

Buklijas, Tatjana. "Culture of Death and the Politics of Corpse Supply: Anatomy in Vienna, 1848–1914." *Bulletin of the History of Medicine* 82, no. 3 (2008): 570–607.

Burns, Jordan N., Rodolfo Acuna-Soto, and David W. Stahle. "Drought and Epidemic Typhus, Central Mexico, 1655–1918." *Historical Review* 20, no. 3 (March 2014): 1.

Bynum, W. F. *Science and the Practice of Medicine in the Nineteenth Century*. New York: Cambridge University Press, 1994.

Cabrejo, Fannie Muñoz. "The New Order: Diversions and Modernization in Turn-of-the-Century Lima." In *Latin American Popular Culture Since Independence*. 2nd ed. Edited by William Beezley and Linda Curcio-Nagy, 153–63. Lanham MD: Rowman and Littlefield, 2012.

Campbell, John A. *A Biographical History with Portraits of Prominent Men of the Great West*. Lincoln NE: Western Biographical and Engraving Company, 1902.

Carregha Lamadrid, Luz. "El Impacto del Ferrocarril en San Luís Potosí durante El Porfiriato." In *Visiones del Porfiriato: Visiones de México*, edited by Jane-Dayle Lloyd, Eduardo N. Mijangos Diaz, Marisa Pérez Domínguez, and María Eugenia Ponce Alcocer, 185–98. Mexico City: Universidad Iberoamericana, 2004.

Carrillo, Ana María. "¿Estado de peste o estado se sitio?: Sinaloa y Baja California, 1902–1903." *Historia Mexicana* 54, no. 4 (April–June 2005): 1049–1103.

———. "Médicos del México decimonónico: entre el control estatal y la autonomía professional." *Dynamis: Acta Hispanica ad Medicinae Scientiarumque Historiam Illustrandam* 22 (2002): 351–75.

———. "Profesiones Sanitarias y Lucha de Poderes en el México del Siglo XIX." *Asclepio* 50, no. 2 (1998): 149–68.

Carter, Eric D. *Enemy in the Blood: Malaria, Environment, and Development in Argentina*. Tuscaloosa: University of Alabama Press, 2012.

Cassity, Brian Ellis. "Health, Sanitation, Hygiene, and Welfare: Public Policy in the Age of the Mexican Revolution." PhD diss., Arizona State University, 2010.

Castronovo, Russ. *Death, Eroticism, and the Public Sphere in the Nineteenth-Century United States*. Durham NC: Duke University Press, 2001.

Cheesman, Oliver D. *Environmental Impacts of Sugar Production*. Cambridge MA: CABI Publishing, 2004.

Chowell, Gerardo, Cécile Viboud, Lone Simonsen, Mark A. Miller, and Rodolfo Acuna-Soto. "Mortality Patterns Associated with the 1918 Influenza in Mexico: Evidence for a Spring Herald Wave and Lack of Pre-Existing Immunity in Older Population." *Journal of Infectious Diseases* 202, no. 4 (August 2010): 567–75.

Coatsworth, John H. *Growth against Development: The Economic Impact of Railroads in Porfirian Mexico*. DeKalb: Northern Illinois University Press, 1981.

Cockcroft, James D. *Intellectual Precursors of the Mexican Revolution, 1900–1913*. Austin: University of Texas Press, 1976.

Codigo Sanitario de los Estados Unidos Mexicanos. 9th ed. Mexico City: Eusebio Sanchez, 1895.

Cohen, Ed. *A Body Worth Defending: Immunity, Biopolitics, and the Apotheosis of the Modern Body*. Durham NC: Duke University Press, 2009.

Cooke, G. "An Extreme Power Engineer—the Accomplishments of Frederick Stark Pearson." *Power and Energy Magazine* 1, no. 6 (November–December 2003): 60–65.

Crosby, Alfred W. *America's Forgotten Pandemic: The Influenza of 1918*. New York: Cambridge University Press, 1989.

Cruz Gómez, Daniel. "El Hospital Juárez." In *Histórica Gráfica de la Medicina Mexicana del Siglo XX*, 228–29. 2nd ed. Mexico City: Méndez Editores, 2003.

Cueto, Marcos. *El regreso de las epidemias: salud y sociedad en el Perú del siglo XX*. Lima: Instituto de Estudios Peruanos, 1997.

———. *Missionaries of Science: The Rockefeller Foundation in Latin America*. Bloomington: Indiana University Press, 1994.

———. *Saberes andinos: ciencia y tecnología en Bolivia, Ecuador, y Perú*. Lima: Instituto de Estudios Peruanos, 1995.

———. *Salud, Cultura y sociedad en América Latina*. Lima: Instituto de Estudios Peruanos, 1996.

Cunningham, Andrew, and Paul Williams, eds. *The Laboratory Revolution in Medicine*. New York: Cambridge University Press, 1992.

Daft, Richard L. *Organization Theory and Design*. 10th ed. Mason OH: South-Western Language Learning, 2008.

David, Paul A. "Path Dependence, Its Critics and the Quest for 'Historical Economics.'" In *Evolution and Path Dependence in Economic Ideas: Past and Present*, edited by Pierre Garrouste and Stavros Ioannides, 15–36. Cheltenham UK: Edward Elgar, 2001.

de la Torre Rendón, Judith. "Las imágenes fotográficas de la sociedad mexicana en la prensa gráfica del Porfiriato." *Historia Mexicana* 48, no. 2 (Octubre–Diciembre 1998): 343–73.

Denzel, Markus A. *Handbook of World Exchange Rates, 1500–1914*. Burlington VT: Ashgate, 2010.

Derrida, Jacques. *Positions*. Translated by Alan Bass. Chicago IL: University of Chicago Press, 1981.

Dufendach, Rebecca Ann. "Injecting Modernity: Regulating Hygiene in Porfirian Oaxaca, Mexico." Master's thesis, Northeastern University, 2008.

Edgerton, David. "From Innovation to Use: Ten Eclectic Theses on the Historiography of Technology." *History and Technology: An International Journal* 16, no. 2 (1999): 111–36.

Esposito, Matthew D. "Death and Disorder in Mexico City: The State Funeral of Manuel Romero Rubio." In *Latin American Popular Culture Since Inde-*

pendence, edited by William H. Beezley and Linda A. Curcio-Nagy, 106–19. New York: Rowman and Littlefield, 2012.

———. *Funerals, Festivals, and Cultural Politics in Porfirian Mexico*. Albuquerque: University of New Mexico Press, 2010.

Estadísticas Históricas de México, vol. 2: Cuadro 21.6. Mexico City: Instituto Nacional de Estadística, Geografía e Informática, 1990.Ewald, Paul W. *Plague Time: The New Germ Theory of Disease*. New York: Anchor Books, 2002.

Farrell, James J. *Inventing the American Way of Death, 1830–1920*. Philadelphia: Temple University, 1980.

Feldenkirchen, Wilfried. *Werner Von Siemens: Inventor and International Entrepreneur*. Columbus: Ohio State University Press, 1994.

Fischer, Martin Henry. *Nephritis: An Experimental and Critical Study of Its Nature and the Principles of Its Relief.* New York: John Wiley and Sons, 1912.

Fitch, Francis E. *The Fitch Record of Government Finances*. 3rd ed. New York: Fitch Publishing, 1918.

Floud, Roderick, and Paul Johnson, eds. *The Cambridge Economic History of Modern Britain: Economic Maturity, 1860–1939*. Vol. 2. New York: Cambridge University Press, 2004.

Forman, Paul. "On the Historical Forms of Knowledge Production and Curation: Modernity Entailed Disciplinarity, Postmodernity Entails Antidisciplinarity." *Osiris* 27, no. 1 (2012): 56–97.

Foucault, Michel. *The Birth of the Clinic: An Archeology of Medical Perception*. Translated by A. M. S. Smith. London: Tavistock, 1976.

———. *Discipline and Punish: The Birth of the Prison*. Translated by Allen Sheridan. London: Allen Lane, Penguin, 1977.

———. *History of Sexuality*. Vol. 1: *An Introduction*. Translated by Robert Hurley. London: Allen Lange, Penguin, 1979.

———. *Madness and Civilization: A History of Insanity in the Age of Reason*. Translated by Richard Howard. New York: Vintage, 1965.

Fowler, Will. *Santa Anna of Mexico*. Lincoln: University of Nebraska Press, 2007.

François, Marie Eileen. *A Culture of Everyday Credit: Housekeeping, Pawn Broking, and Governance in Mexico City, 1750–1920*. Lincoln: University of Nebraska Press, 2005.

"Frederick Pearson Starks." *Successful American*, February 1903: 92–94.

French, William. *A Peaceful and Working People: Manners, Morals, and Class Formation in Northern Mexico*. Albuquerque: University of New Mexico, 2008.

———. "Progreso Forzado: Workers and the Inculcation of the Capitalist Work Ethic in the Parral Mining District." In *Rituals of Rule, Rituals of Resistance: Public Celebrations and Popular Culture in Mexico*, edited by William H. Beezley, Cheryl E. Martin, and William E. French, 191–207. Wilmington DE: Scholarly Resources, 1994.

Fuller, Elmer Dean. *Handbook of the Law of Mexican Commercial Corporations*. Mexico City: Mexican Law-Book Publishing, 1911.

"Funeral Cars in Mexico." *Electric Railway Journal* 14, no. 3 (March 1898): 131–32.

Gallo, Rubén. *Mexican Modernity: The Avant-Garde and the Technological Revolution*. Cambridge MA: MIT Press, 2005.

García, Víctor, and Laura González. "Juramentos and Mandas: Traditional Catholic Practices and Substance Abuse in Mexican Communities of Southeastern Pennsylvania." In *Invisible Anthropologies: Engaged Anthropology in Immigrant Communities*, edited by Alayne Unterberger, 47–63. Hoboken NJ: Wiley-Blackwell, 1999.

García-Ayluardo, Clara. "Confraternity, Cult, and Crown in Mexico City, 1700–1810." PhD diss., University of Cambridge, 1989.

García Héras, Raul. *Transportes, negocios y política la Compañía Anglo Argentina de Tranvías, 1876–1981*. Buenos Aires: Sudamericana, 1994.

Garza, James. *The Imagined Underworld: Sex, Crime, and Vice in Porfirian Mexico City*. Lincoln: University of Nebraska Press, 2007.

Garza, Miguel Z. "¿Es la incineración de cadáveres anticristiano?" El Abogado Cristiano Ilustrado. July 15, 1909.

Geison, Gerald. "'Divided We Stand': Physiologists and Clinicians in the American Context." In *The Therapeutic Revolution: Essays in the Social History of American Medicine*, edited by Morris J. Vogel and Charles E. Rosenberg, 67–90. Philadelphia: University of Pennsylvania Press, 1979.

Gerschenkron, Alexander. *Economic Backwardness in Historical Perspectives: A Book of Essays*. Cambridge MA: Belknap Press, 1962.

Giddens, Anthony. *Modernity and Self-Identity: Self and Society in the Late Modern Age*. Palo Alto CA: Stanford University Press, 1994.

Giedion, Sigfried. *Space, Time, and Architecture: The Growth of a New Tradition*. 4th ed. Boston: Harvard University Press, 1962.

Gilpin Faust, Drew. *This Republic of Suffering: Death and the American Civil War*. New York: Vintage Books, 2009.

González Navarro, Moisés. *Estadisticas Sociales del Porfiriato, 1877–1910*. Mexico: Talleres Graficos de la Nacion, 1956.

Grandin, Greg. *Fordlandia: The Rise and Fall of Henry Ford's Forgotten Jungle City*. New York: Metropolitan Books, 2009.

Habenstein, Robert W., and William M. Lamers. *The History of American Funeral Directing*. 6th ed. Brookfield WI: National Funeral Directors Association, 2007.

Haber, Steven. *Efficiency and Uplift*. Chicago IL: University Press of Chicago, 1964.

———. *Industry and Underdevelopment: The Industrialization of Mexico, 1890–1940*. Palo Alto CA: Stanford University Press, 1989.

Haber, Stephen, Armando Razo, and Noel Maurer. *Politics of Property Rights: Political Instability, Credible Commitments, and Economic Growth in Mexico, 1878–1929*. New York: Cambridge University Press, 2003.

Hale, Charles A. *Mexican Liberalism in the Age of Mora, 1821–1853*. New Haven CT: Yale University Press, 1968.

———. *The Transformation of Liberalism in Late-Nineteenth Century Mexico*. Princeton NJ: Princeton University Press, 1989.

Haller, John S., Jr. *Medical Protestants: The Eclectic American Medicine, 1825–1939*. Carbondale: Southern Illinois University, 1994.

Hardwicke, Herbert J. *Medical Education and Practice in All Parts of the World*. Philadelphia: Presley Blakiston, 1880.

Harrison, Mark. *Public Health in British India: Anglo-Indian Preventive Medicine, 1855–1914*. New York: Cambridge University Press, 1994.

Hart, John Mason. *Empire and Revolution: The Americans in Mexico since the Civil War*. Berkeley: University of California Press, 2002.

———. *Revolutionary Mexico: The Coming and Process of the Mexican Revolution*. Berkeley: University of California Press, 1987.

Hassig, Ross. *Mexico and the Spanish Conquest*. 2nd ed. Norman: University of Oklahoma Press, 2006.

Hernández, Luz María, and George M. Foster. "Curers and Their Cures in Colonial New Spain and Guatemala: The Spanish Component." In *Mesoamerican Healers*, edited by Brad R. Huber and Alan R. Sandstrom, 19–46. Austin: Texas University Press, 2001.

Hohenschuh, W. P. *The Modern Funeral: Its Management*. Chicago: Trade Periodical, 1900.

Hunt, F. W. *The Homeopathic Theory and Practice of Medicine*. Vol. 1. New York: William Radde, 1868.

Hunter, Michael. *Robert Boyle, 1627–1691: Scrupulosity and Science*. Rochester NY: Boydell Press, 2000.

———, ed. *Robert Boyle Reconsidered*. New York: Cambridge University Press, 1994.

Huntington, George S. "The Morphological Museum as an Education Factor in the University System." *Science* 13, no. 329 (April 1, 1901): 601–11.

Iglesia, Ramón. *Columbus, Cortés, and Other Essays*. Translated by Lesley Simpson. Berkeley: University of California Press, 1969.

Iserson, Kenneth. *Death to Dust: What Happens to Dead Bodies?* Tuscon AZ: Galen Press, 1994.

Janvier, Thomas A. *The Mexican Guide*. New York: Charles Scribner's Sons, 1886.

Jeremy, David J. *Transatlantic Industrial Revolution: The Diffusion of Textile Technologies between Britain and America, 1790–1830s*. Oxford: Basil Blackwell, 1981.

Johns, Michael. *The City of Mexico in the Age of Díaz*. Austin: University of Texas Press, 1997.

Johnson, Lyman L., ed. *Death, Dismemberment, and Memory: Body Politics in Latin America*. Albuquerque: University of New Mexico Press, 2004.

———. *The Problem of Order in Changing Societies: Essays in Crime and Policing in Argentina and Uruguay, 1750–1919*. Albuquerque: University of New Mexico Press, 1990.

Joseph, Gilbert M., and Daniel Nugent, ed. *Everyday Forms of State Formation: Revolution and the Negotiation of Rule in Modern Mexico*. Durham NC: Duke University Press, 1994.

Kalach, Alberto. "Architecture and Place: The Stadium of the University City." In *Modernity and the Architecture of Mexico*, edited by Edward Rudolf Burian, 107–15. Austin: University of Texas Press, 1997.

Kanigel, Robert. *The One Best Way: Frederick Winslow Taylor and the Enigma of Efficiency*. New York: Viking Penguin, 1997.

Kapelusz-Poppi, Ana María. "Provincial Intellectuals from Michoacán and the Professionalization of the Post-Revolutionary Mexican State." PhD diss., University of Illinois at Chicago, 2002.

———. "Rural Health and State Construction in Post-Revolutionary Mexico: The Nicolaita Project for Rural Medical Services." *The Americas* 58, no. 2 (October 2001): 261–83.

Kelly, Patty. "Into the Galactic Zone: Managing Sexuality in Neoliberal Mexico." In *Policing Pleasure: Sex Work, Policy, and the State in a Global Perspective*, edited by Susan Dewey and Patty Kelly, 31–44. New York: New York University Press, 2011.

Kemper, Robert V., and Anya Peterson Royce. "Mexico Urbanization since 1821: A Macro-Historical Approach." *Urban Anthropology* 8, no. 3 (Winter 1979): 267–89.

———. "Urbanization in Mexico: Beyond the Heritage of the Conquest." In *Heritage of Conquest: Thirty Years Later*, edited by Carl Kendall, John Hawkins, and Laurel Bassell, 92–128. Albuquerque: University of New Mexico Press, 1983.

Khan, B. Zorina. *The Democratization of Invention: Patents and Copyrights in American Economic Development, 1790–1920*. New York: Cambridge University Press, 2005.

Kirkwood, J. Burton. *The History of Mexico*. 2nd ed. Santa Barbara CA: Greenwood Press, 2010.

Knight, Alan. *The Mexican Revolution*. 2 vols. Lincoln: University of Nebraska Press, 1986.

Knox, Frederick John. *The Anatomist's Instructor, and Museum Companion: Being Practical Directions for the Formation and Subsequent Management of Anatomical Museums*. Edinburgh: Charles Black, 1836.

Kong, Lily. Foreword to *Deathscapes: Spaces for Death, Dying, Mourning, and Remembrance*, edited by Avril Maddrell and James D. Sidaway, xv–xvii. Burlington VT: Ashgate, 2010.

LaBerge, Ann F. *Mission and Method: The Early Nineteenth-Century French Public Health Movement*. New York: Cambridge University Press, 1992.

Larkin, Brian. "Confraternities and Community: The Decline of the Communal Quest for Salvation in Eighteenth-Century Mexico City." In *Local Religion in Colonial Mexico*, edited by Martin Nesvig, 189–214. Albuquerque: University of New Mexico Press, 2006.

Latour, Bruno. *We Have Never Been Modern*. Translated by Catherine Porter. Cambridge MA: Harvard University Press, 1993.

Lavery, Brian. *Horatio Lord Nelson*. New York: New York University Press, 2003.
Lear, John R. "Mexico City: Space and Class in the Porfirian Capital, 1884–1910." *Journal of Urban History* 22, no. 4 (May 1996): 444–92.
———. *Workers, Neighbors, and Citizens: The Revolution in Mexico City*. Lincoln: University of Nebraska Press, 2001.
Leiby, John S. "The Royal Indian Hospital of Mexico City, 1553–1680." *The Historian* 57, no. 3 (1995): 573–80.
Lerman-Garber, Israel, Francisco J. Gómez-Pérez, and Ricardo Quibrera-Infante. "Mexico." In *The Epidemiology of Diabetes Mellitus: An International Perspective*, edited by Jean-Marie Ekoé, Paul Zimmet, and Rhys Williams, 195–204. New York: John Wiley and Sons, 2001.
Lewis, Oscar. *Tepoztlán: Village in Mexico*. New York: Holt, Rinehart and Winston, 1960.
Licéaga, Eduardo. *Memoria de los trabajos realizados en la escuela de medicina por el año 1907*. Mexico City: Imprenta de A. Carranza y Compañía, 1908.
Lomelí Vanegas, Leonardo. "Ciencia Económico y Positivismo: Hacia una nueva interpretación de la política económica del Porfiriato." In *Visiones del Porfiriato: Visiones de México*, edited by Jane-Dayle Lloyd, Eduardo N. Mijangos Diaz, Marisa Pérez Domínguez, and María Eugenia Ponce Alcocer, 199–222. Mexico City: Universidad Iberoamericana, 2004.
Lomnitz, Claudio. *Death and Dying in Mexico*. New York: Zone Books, 2005.
López, Amanda M. "The Cadaverous City: The Everyday Life of the Dead in Mexico City, 1875–1930." PhD diss., University of Arizona, 2010.
———. "'An Urgent Need for Hygiene': Cremation, Class, and Public Health in Mexico City, 1879–1920." *Mexican Studies/Estudios Mexicanos* 31, no. 1 (2015): 88–124.
López-Alonso, Moramay. *Measuring Up: A History of Living Standards in Mexico, 1850–1950*. Palo Alto CA: Stanford University Press, 2012.
López Ramos, Sergio. *History of the Air and Other Smells in Mexico City, 1840–1900*. Bloomington IN: Palibrio, 2016.
MacDonald, Helen. *Human Remains: Episodes in Human Dissection*. Melbourne, Australia: Melbourne University Press, 2005.
Mallon, Florencia E. "The Promise and Dilemma of Subaltern Studies: Perspectives from Latin American History." *American Historical Review* 99, no. 5 (December 1994): 1491–515.
Marroquí, José María. *La ciudad de México*. Mexico City: Tip y Lit La Europea, 1900.
Martin, Percy Falcke. *Mexico of the Twentieth Century*. Vol. 1. London: Edward Arnold, 1907.
Matthews, Michael D. *The Civilizing Machine: A Cultural History of Mexican Railroads, 1876–1910*. Lincoln: University of Nebraska Press, 2013.
———. "De Viaje: Elite Views of Modernity and the Porfirian Railway Boom." *Estudios Mexicanos* 26, no. 2 (Summer 2010): 251–89.
———. "Railway Culture and the Civilizing Mission in Mexico, 1876–1910." PhD diss., University of Arizona, 2008.

Mayer, R. G. *Embalming: History, Theory, and Practice*. Norwalk CT: Appleton and Lange, 1990.
McCaa, Robert. "The Peopling of 19th Century Mexico: Critical Scrutiny of a Censured Century." *Statistical Abstract of Latin America* 30 (1993): 606.
McCrea, Heather. *Diseased Relations: Epidemics, Public Health, and State-Building in Yucatán, Mexico, 1847–1924*. Albuquerque: University of New Mexico Press, 2010.
McKeown, Thomas. *The Modern Rise of Population*. New York: Academic Press, 1976.
McKeown, Thomas, and R. G. Brown. "Medical Evidence Related to English Population Changes in the Eighteenth Century." *Population Studies* 9, no. 2 (November 1955): 119–41.
McKeown, Thomas, R. G. Brown, and R. G. Record. "An Interpretation of the Modern Rise of Population in Europe." *Population Studies* 26, no. 3 (April 1972): 345–82.
McKeown, Thomas, and R. G. Record. "Reasons for the Decline of Mortality in England and Wales during the Nineteenth Century." *Population Studies* 16, no. 2 (November 1962): 94–122.
McKeown, Thomas, R. G. Record, and R. D. Turner. "An Interpretation of the Decline of Mortality in England and Wales during the Twentieth Century." *Population Studies* 29, no. 3 (November 1975): 391–422.
McKiernan-González, John. *Fevered Measures: Public Health and Race at the Texas-Mexico Border, 1848–1942*. Durham NC: Duke University Press, 2012.
Megged, Amos. *Social Memory in Ancient and Colonial Mesoamerica*. New York: Cambridge University Press, 2010.
Miles, George, ed. *World's Fairs: From London 1851 to Chicago, IL 1893*. Chicago: Midway Publishing, 1892.
Millet, Martin. *Roman Britain*. London: B. T. Betsford Limited, 1995.
"The Mining States of Mexico." *Overland Monthly* 56 (July–December 1910): 64.
"Miscellany." *Journal of the American Medical Association* 29 (August 7, 1897): 298.
Mitchell, Timothy J. *Intoxicating Identities: Alcohol's Power in Mexican History and Culture*. New York: Routledge, 2004.
Mitford, Jessica. *American Way of Death*. New York: Simon and Schuster, 1963.
Mokyr, Joel. *The Levers of Riches: Technological Creativity and Economic Progress*. New York: Oxford University Press, 1990.
Moreno-Brid, Juan Carlos, and Jaime Ros. *Development and Growth in the Mexican Economy: A Historical Perspective*. New York: Oxford University Press, 2009.
Mraz, John. *Looking for Mexico: Modern Visual Culture and National Identity*. Durham NC: Duke University Press, 2009.
Nas, Peter J. M., and Pierpaulo De Giosa. "Conclusion: Feeling at Home in the City and the Codification of Urban Symbolism Research." In *Cities Full of Symbols: A Theory of Urban Space and Culture*, edited by Peter J. M. Nas, 283–92. Amsterdam: Leiden University Press, 2011.

Nelson, Daniel. *Frederick W. Taylor and the Rise of Scientific Management*. Madison: University of Wisconsin Press, 1980.

Nelson, Wolfred. "Yellow Fever." In *Twentieth Century Practice: An International Encyclopedia of Modern Medical Science*, edited by Thomas Lathrop Stedman, 397. New York: William Wood, 1900.

"Obituaries." *American Medicine* 7, no. 14 (April 2, 1904): 540–41.

Olsen, Patrice Elizabeth. *Artifacts of Revolution: Architecture, Society and Politics in Mexico City, 1920–1940*. London: Rowman and Littlefield, 2008.

O'Rourke, Kevin H., and Jeffrey G. Wilson. *Globalization and History: The Evolution of a Nineteenth-Century Atlantic Economy*. Cambridge MA: MIT Press, 1999.

Overmeyer-Velázquez, Mark. *Visions of the Emerald City: Modernity, Tradition, and the Formation of Porfirian Oaxaca, Mexico*. Durham NC: Duke University Press, 2006.

Palmer, Steven. *From Popular Medicine to Medical Populism: Doctors, Healers and Public Power in Costa Rica, 1800–1940*. Durham NC: Duke University Press, 2003.

Peard, Julyan. *Race, Place, and Medicine: The Idea of the Tropics in Nineteenth-Century Brazil*. Durham NC: Duke University Press, 2000.

Pérez-Rayón, Nora. "La Sociología de lo cotidiano: Discursos y fiestas cívicas en el México de 1900." *Sociológica* 8, no. 23 (1993): 171–98.

Piccato, Pablo. *City of Suspects: Crime in Mexico City, 1900–1931*. Durham NC: Duke University Press, 2001.

———. "'El Chalequero' or the Mexican Jack the Ripper: The Meanings of Sexual Violence in Turn-of-the-Century Mexico City." *Hispanic American Historical Review* 81, nos. 3–4 (August–November 2001): 623–51.

Pilcher, Jeffrey. "Abattoir or Packinghouse?: A Bloody Industrial Dilemma in Mexico City, c. 1890." In *Meat, Modernity, and the Rise of the Slaughterhouse*, edited by Paula Young Lee, 216–36. Lebanon: University of New Hampshire Press, 2008.

———. *The Sausage Rebellion: Public Health, Private Enterprise, and Meat in Mexico City, 1890–1917*. Albuquerque: University of New Mexico Press, 2006.

Pitt, Leonard. *Walks through Lost Paris: A Journey into the Heart of Historic Paris*. Washington DC: Shoemaker and Hoard, 2006.

Porro, Alessandro, Bruno Falconi, Carlo Cristini, Lorenzo Lorusso, and Antonia F. Franchini. "Modernity in Medicine and Hygiene at the End of the 19th Century: The Example of Cremation." *Journal of Public Health Research* 1, no. 1 (2012): 51–58.

Porter, Roy. *Disease, Medicine and Society in England, 1550–1860*. 2nd ed. New York: Cambridge University Press, 1995.

Potter, William Warren. "American Public Health Association: A Historical Sketch—History of the Preparation for the Buffalo Meeting, 1896." *Buffalo Medical Journal* 36, no. 2 (August 1896–July 1897): 114.

Powell, T. G. "Priest and Peasants in Central Mexico: Social Conflict during 'La Reforma.'" *Hispanic American Historical Review* 57, no. 2 (May 1977): 296–313.

Prakash, Gyan. "Subaltern Studies as Postcolonial Criticism." *American Historical Review* 99, no. 5 (December 1994): 1475–90.

Prantl, Adolfo, and José L. Grosó. *La ciudad de México: novísima guía universal de la capital de la República Mexicana*. Mexico City: Juan Buxó y Compañía editores, Librería Madrileña, 1901.

Prothero, Stephen. *Purified by Fire: A History of Cremation in America.* Berkeley: University of California Press, 2001.

"Railroading in Mexico." *Railway Agent and Station Agent: A Monthly Journal Devote to the Interests of Local Freight and Freight Agents* 1 (March 1889): 172–73.

Ramírez de Arrellano, Nicolás. "Higiene: profilaxis de la rabia." *Gaceta Médica de México* 24 (June 1, 1889): 206–9.

Redfield, Robert, and Margaret Redfield. *Disease and Its Treatment in Dzitas, Yucatan*. Washington DC: Carnegie Institution of Washington, 1940.

Reese, Thomas F., and Carol McMichael Reese. "Revolutionary Urban Legacies: Porfirio Díaz's Celebrations of the Centennial of Mexican Independence in 1910." In *Arte, Historia e Identidad en América: Visiones Comparativas; XVII Coloquio Internacional de Historia del Arte*, Vol. 2, edited by Renato González Mello and Juana Gutiérrez Haces, 361–73. Mexico City: Universidad Nacional Autónoma de México, 1994.

Richardson, Ruth. *Death, Dissection and the Destitute*. London: Routledge Press, 1987.

Rivera Cambas, Manuel. *Mexico pintoresco, artistico y monumental*. 2nd ed. Mexico City: Imprenta de la Reforma, 1882.

Rivera Garza, Cristina. "Masters of the Street: Bodies, Power, and Modernity in Mexico, 1867–1930." PhD diss., University of Houston, 1995.

Robbins, Louise E. *Louis Pasteur and the Hidden World of Microbes.* New York: Oxford University Press, 2001.

Robinson, James A., and Thierry Verdier. "The Political Economy of Clientelism." *Scandinavian Journal of Economics* 115, no. 2 (2013): 260–91.

Rogaski, Ruth. *Hygienic Modernity: Meaning of Health and Disease in Treaty-Port China*. Berkeley: University of California Press, 2004.

Rohé, George H. "Resolutions Adopted by the American Public Health Association Relative to the Bertillon Classification of Causes of Death." *32nd Annual Report of the Secretary of State on the Registration of Births and Deaths, Marriages and Divorces in Michigan for the Year 1898*. Lansing MI: Robert Smith Printing, 1900.

Rohlfes, Laurence J. "Police and Penal Correction in Mexico City, 1876–1911: A Study of Order and Progress in Porfirian Mexico." PhD diss., Tulane University, 1983.

Rosen, George. *History of Public Health*. New York: MD Publications, 1958.

Rosenberg, Charles. *The Cholera Years: The United States in 1832, 1849, and 1866*. Chicago: University of Chicago Press, 1987.

Rosenberg, Nathan. "Economic Development and the Transfer of Technology: Some Historical Perspectives." *Technology and Culture* 11, no. 4 (October 1970): 550–75.

Rosenthal, Anton. "The Arrival of the Electric Streetcar and the Conflict over Progress in Early Twentieth-Century Montevideo." *Journal of Latin American Studies* 27, no. 2 (May 1995): 319–41.

Ross, Paul. "Mexico's Superior Health Council and the American Public Health Association: The Transnational Archive of Porfirian Public Health, 1887–1910." *Hispanic American Historical Review* 89, no. 4 (November 2009): 573–603.

———. "From Sanitary Police to Sanitary Dictatorship: Mexico's 19th Century Public Health Movement." PhD diss., University of Chicago, 2005.

Rothstein, William G. *American Medical Schools and the Practice of Medicine*. New York: Oxford University Press, 1987.

Rubenstein, Anne. *Bad Language, Naked Ladies, and Other Threats to the Nation: A Political History of Comic Books in Mexico*. Durham NC: Duke University Press, 1998.

Ruggiero, Kristin. *Modernity in the Flesh: Medicine, Law, and Society in Turn-of-the-Century Argentina*. Palo Alto CA: Stanford University Press, 2004.

Ruiz-Alfaro, Sofia. "A Threat to the Nation: México Marimacho and Female Masculinities in Postrevolutionary Mexico." *Hispanic American Historical Review* 81, no. 1 (Winter 2013): 41–62.

Rydell, Robert W. *World of Fairs: The Century-of-Progress Expositions*. Chicago: University of Chicago Press, 1993.

Sabel, C. F. *Work and Politics: The Division of Labor in Industry*. New York: Cambridge University Press, 1982.

Sánchez, Adriana Alvárez. "La Universidad Nacional de México y El Centenario de la Independencia." Paper presented at XIV Encuentro de Latinoamericanistas Españoles, Universidad de Santiago de Compostela, Galicia, Spain, September 15–18, 2010.

Sandwick, Richard L. "More Men in Public School." *Popular Science Monthly* 65 (September 1904): 443–51.

Sappol, Michael. "'Morbid Curiosity': The Decline and Fall of the Popular Anatomical Museum." *Common-Place* 4, no. 2 (January 2004). http:// www.common-place.org/vol-04/no-02/sappol, accessed August 8, 2012.

———. *A Traffic of Dead Bodies: Anatomy and Embodied Social Identity in Nineteenth Century America*. Princeton NJ: Princeton University Press, 2002.

Schell, Patience. *Church and State Education in Revolutionary Mexico City*. Tuscon: University of Arizona Press, 2003.

Scott, James C. *The Moral Economy of the Peasant: Rebellion and Subsistence in Southeast Asia*. New Haven CT: Yale University Press, 1976.

———. *Seeing Like a State: How Certain Schemes to Improve the Human Condition Have Failed*. New Haven CT: Yale University, 1998.

———. *Weapons of the Weak: Everyday Forms of Peasant Resistance*. New Haven CT: Yale University Press, 1985.

Shah, Nayan. *Contagious Divides: Epidemics and Race in San Francisco's Chinatown*. Berkeley: University of California Press, 2001.

Shattuck, George C. *The Peninsula of Yucatan: Medical, Biological, Meteorological and Sociological Studies*. Washington DC: Carnegie Institution of Washington, 1933.

Shortt, S. E. D. "Physicians, Science, and Status: Issues in the Professionalization of Anglo-American Medicine in the Nineteenth-Century." *Medical History* 27 (1983): 51–68.

Shryock, Richard. *American Medical Research, Past and Present*. New York: Commonwealth Fund, 1947.

Sigerist, Henry E. *The Great Doctors: A Biographical History of Medicine*. New York: Norton, 1933.

Singher, Martial, and Eta Singher. *An Interpretative Guide of Operatic Arias: A Handbook for Singers, Coaches, Teachers, and Students*. University Park: Pennsylvania State University Press, 1983.

Sloan, Kathryn. *Death in the City: Suicide and the Social Imaginary in Modern Mexico*. Berkeley: University of California Press, 2017.

Smith, Merritt Roe. "The Political Economy of Pacing." In *Major Problems in the History of American Technology: Document and Essays*, edited by Merritt Roe Smith and Gregory Clancey, 182–90. New York: Houghton Mifflin, 1997.

"Society Proceedings." *Buffalo Medical Journal* 36, no. 2 (August 1896–July 1897): 203.

Söderquist, Thomas, ed. *The Historiography of Contemporary Science and Technology*. Amsterdam: Harwood Academic, 1997.

Sowell, David. "Quacks and Doctors: The Construction of Biomedical Authority in Mexico." *Juanita Voices* 5 (2005): 15.

———. *The Tale of Healer Miguel Perdomo Neira: Medicine, Ideologies, and Power in the Nineteenth-Century Andes*. Wilmington DE: Scholarly Resources, 2001.

Spivak, Gayatri. "Can the Subaltern Speak?" In *Marxism and the Interpretation of Culture*, edited by Cary Nelson and Lawrence Grossberg, 271–316. Urbana: University of Illinois Press, 1988.

Staples, Anne. "Policia y Buen Gobierno: Municipal Efforts to Regulate Public Behavior, 1821–1857." In *Rituals of Rule, Rituals of Resistance: Public Celebrations and Popular Culture in Mexico*, edited by William H. Beezley, Cheryl E. Martin, and William E. French, 115–26. Wilmington DE: Scholarly Resources, 1994.

Starr, Frederick. *Readings from Modern Mexican Authors*. Chicago IL: Open Court, 1904.

Starr, Paul. *The Social Transformation of American Medicine*. New York: Basic Books, 1982.

Sturgis, Russell. *A Dictionary of Architecture and Building: Biographical, Historical and Descriptive*. Vol. 2. New York: MacMillan Company, 1905.

Sullivan, B. G. "The Challenge of Economic Transformation." In *Technological Change and the Transformation of America*, edited by S. E. Goldberg and C. R. Strain, 91–103. Carbondale: Southern Illinois University Press, 1987.

Suprema Sacra Congregatio Sancti Officii. *Instructio Piam et constantem: de cadaverum cremation*. *Acta Apostolicae Sedis* 56 (1964): 822–24.

Tate Lenning, John. *The Royal Protomedicato: The Regulation of the Medical Professions in the Spanish Empire*, edited by John Jay TePaske. Durham NC: Duke University Press, 1985.

Taylor, Frederick W. *Principles of Scientific Management*. New York: Harper Brothers, 1911.

Taylor, Timothy. *The Buried Soul: How Humans Invented Death*. Boston: Beacon Press, 2002.

Tenenbaum, Barbara. "Streetwise History: The Paseo de la Reforma and the Porfirian State, 1876–1910." In *Rituals of Rule, Rituals of Resistance: Public Celebrations and Popular Culture in Mexico*, edited by William H. Beezley, Cheryl E. Martin, and William E. French, 127–50. Wilmington DE: Scholarly Resources, 1994.

Tenorio-Trillo, Mauricio. *I Speak of the City: Mexico City at the Turn of the Twentieth Century*. Chicago: University of Chicago Press, 2013.

———. *Mexico at the World's Fairs: Crafting a Modern Nation*. Berkeley: University of California Press, 1996.

———. "1910 Mexico City: Space and Nation in the City of the Centenario." *Journal of Latin American Studies* 28, no.1 (February 1996): 75–104.

Terry, T. Philip. *Mexico: An Outline Sketch of the Country, Its People, and Their History from the Earliest Times to the Present*. New York: Houghton Mifflin, 1914.

———. *Terry's Mexico: Handbook for Travellers*. 2nd ed. New York: Houghton Mifflin, 1911.

Thompson, E. P. *The Making of the English Working Class*. New York: Pantheon Books, 1964.

———. "Time, Work-Discipline, and Industrial Capitalism." *Past and Present* 38 (December 1967): 56–97.

Toledo Martínez, Hayde Yazmín. "Historia social de la tecnología tranviaria en el Distrito Federal, 1898–1920." Master's thesis, Universidad Nacional Autónoma de México, 2010.

Toxqui Garay, María Áurea. "'El Recreo de Los Amigos': Mexico City's Pulquerías during the Liberal Republic (1856–1911)." PhD diss., University of Arizona, 2008.

Trexler, Richard C. *Reliving Golgotha: The Passion Play of Iztapalapa*. Boston: Harvard University Press, 2003.

Tulchinsky, Theodore H., and Elena D. Varivakova. *The New Public Health*. 2nd ed. Burlington VT: Elsevier Academic, 2009.

Tweedie, Ethel. *Mexico As I Saw It*. London: Hurst and Blackett, 1901.

———. *Porfirio Díaz: Seven Times President*. London: Hurst and Blackett, 1906.

Urbain, Jean-Didier. *La Societe de Conservation: Étude Sémiologique des Cimitières d'Occident*. Paris: Payot, 1978.

Van Der Bent, Teunis J. *The Problem of Hygiene in Man's Dwellings: A Textbook for Students of Architecture, Household Arts, Practical Arts and Hygiene of Private and Institutional Dwellings*. New York: Architectural Book Publishing Company, 1920.

Van Hoy, Teresa Miriam. *A Social History of Mexico's Railroads: Peons, Prisoners, and Priests*. London: Rowman and Littlefield, 2008.

Vega, R. A., and Ortega Baez. "La Óptica Metodista en la Divulgación de la Medicina Científica: El Abogado Cristiano Ilustrado, 1877–1910." *Eä Journal* 1, no. 2 (December 2009): 1–25.

Verdery, Katherine. *The Political Lives of Dead Bodies: Reburial and Post-Socialist Change*. New York: Columbia University Press, 1999.

Viquiera-Alban, Juan Pedro. "El sentimiento de la muerte en el México ilustrado del siglo XVIII a través de dos textos de la época." *Relaciones* 2, no. 5 (Winter 1981): 27–63.

Voekel, Pamela. *Alone before God: The Religious Origins of Modernity in Mexico*. Durham NC: Duke University Press, 2002.

———. "Peeing on the Palace: Bodily Resistance to Bourbon Reforms in Mexico City." *Journal of Historical Sociology* 5, no. 2 (June 1992): 183–208.

Warner, John Harley. "The Histories of Science and the Science in Medicine." *Osiris* 10 (1995): 164–93.

———. "The Science in Medicine." *Osiris* 1 (1985): 37–58.

Warner, John Harley, and James M. Edmonson. *Dissection: Photographs of a Rite of Passage in American Medicine, 1880–1930*. New York: Blast Books, 2009.

Wasserman, Mark. *Everyday Life and Politics in Nineteenth Century Mexico: Men, Women, and War*. Albuquerque: University of New Mexico Press, 2000.

Wells, Stéphanie. "The Political Development of the National Autonomous University of Mexico." PhD diss., University of Chicago, 1970.

Whitmore, Sylvia D. "Lord Kingsborough and His Contribution to Ancient Mesoamerican Scholarship: The Antiquities of Mexico." *PARI Journal* 9, no. 4 (Spring 2009): 8–16.

Whooley, Owen. *Knowledge in the Time of Cholera: The Struggle over American Medicine in the Nineteenth Century*. Chicago: University of Chicago Press, 2013.

Williams, Raymond. *Culture and Materialism: Selected Essays*. New York: Verso Books, 1980.

Woodhead, G. Sims. *Practical Pathology: A Manual for Students and Practitioners*. 3rd ed. Philadelphia: J. B. Lipincott Coy, 1892.

Zayas Enríquez, Rafael. *Porfirio Díaz*. New York: D. Appleton, 1908.

Zepeda, Tomás. *La Educación Pública en la Nueva España en el Siglo XVI*. Mexico City: Editorial Progreso, 1993.

Zulawski, Ann. "Hygiene and the 'Indian Problem': Ethnicity and Medicine in Bolivia, 1910–1920." *Latin American Research Review* 35, no. 2 (Spring 2000): 107–29.

———. *Unequal Cures: Public Health and Political Change in Bolivia, 1900–1950*. Durham NC: Duke University Press, 2007.

INDEX

In The Mexican Experience series

Women Made Visible: Feminist Art and Media in Post-1968 Mexico City
Gabriela Aceves Sepúlveda

From Idols to Antiquity: Forging the National Museum of Mexico
Miruna Achim

Seen and Heard in Mexico: Children and Revolutionary Cultural Nationalism
Elena Jackson Albarrán

Railroad Radicals in Cold War Mexico: Gender, Class, and Memory
Robert F. Alegre
Foreword by Elena Poniatowska

Mexicans in Revolution, 1910–1946: An Introduction
William H. Beezley and Colin M. MacLachlan

Routes of Compromise: Building Roads and Shaping the Nation in Mexico, 1917–1952
Michael K. Bess

Apostle of Progress: Modesto C. Rolland, Global Progressivism, and the Engineering of Revolutionary Mexico
J. Justin Castro

Radio in Revolution: Wireless Technology and State Power in Mexico, 1897–1938
J. Justin Castro

San Miguel de Allende: Mexicans, Foreigners, and the Making of a World Heritage Site
Lisa Pinley Covert

Celebrating Insurrection: The Commemoration and Representation of the Nineteenth-Century Mexican Pronunciamiento
Edited and with an introduction by Will Fowler

Forceful Negotiations: The Origins of the Pronunciamiento *in Nineteenth-Century Mexico*
Edited and with an introduction by Will Fowler

Independent Mexico: The Pronunciamiento *in the Age of Santa Anna, 1821–1858*
Will Fowler

Malcontents, Rebels, and Pronunciados*: The Politics of Insurrection in Nineteenth-Century Mexico*
Edited and with an introduction by Will Fowler

Working Women, Entrepreneurs, and the Mexican Revolution: The Coffee Culture of Córdoba, Veracruz
Heather Fowler-Salamini

The Heart in the Glass Jar: Love Letters, Bodies, and the Law in Mexico
William E. French

"Muy buenas noches": Mexico, Television, and the Cold War
Celeste González de Bustamante
Foreword by Richard Cole

The Plan de San Diego: Tejano Rebellion, Mexican Intrigue
Charles H. Harris III and Louis R. Sadler

The Inevitable Bandstand: The State Band of Oaxaca and the Politics of Sound
Charles V. Heath

Redeeming the Revolution: The State and Organized Labor in Post-Tlatelolco Mexico
Joseph U. Lenti

Gender and the Negotiation of Daily Life in Mexico, 1750–1856
Sonya Lipsett-Rivera

Mexico's Crucial Century, 1810–1910: An Introduction
Colin M. MacLachlan and William H. Beezley

The Civilizing Machine: A Cultural History of Mexican Railroads, 1876–1910
Michael Matthews

Street Democracy: Vendors, Violence, and Public Space in Late Twentieth-Century Mexico
Sandra C. Mendiola García

The Lawyer of the Church: Bishop Clemente de Jesús Munguía and the Clerical Response to the Liberal Revolution in Mexico
Pablo Mijangos y González

From Angel to Office Worker: Middle-Class Identity and Female Consciousness in Mexico, 1890–1950
Susie S. Porter

¡México, la patria! Propaganda and Production during World War II
Monica A. Rankin

A Revolution Unfinished: The Chegomista Rebellion and the Limits of Revolutionary Democracy in Juchitán, Oaxaca
Colby Ristow

Murder and Counterrevolution in Mexico: The Eyewitness Account of German Ambassador Paul von Hintze, 1912–1914
Edited and with an introduction by Friedrich E. Schuler

Deco Body, Deco City: Female Spectacle and Modernity in Mexico City, 1900–1939
Ageeth Sluis

Pistoleros and Popular Movements: The Politics of State Formation in Postrevolutionary Oaxaca
Benjamin T. Smith

Alcohol and Nationhood in Nineteenth-Century Mexico
Deborah Toner

Death Is All around Us: Corpses, Chaos, and Public Health in Porfirian Mexico City
Jonathan M. Weber

To order or obtain more information on these or other University of Nebraska Press titles, visit nebraskapress.unl.edu.

CPSIA information can be obtained
at www.ICGtesting.com
Printed in the USA
LVHW042239070723
751729LV00002B/222